Sheet Metal Meso- and Microforming and Their Industrial Applications

Sheet Metal Meso- and Microforming and Their Industrial Applications

Xin Min Lai, Ming Wang Fu, and
Lin Fa Peng

CRC Press
Taylor & Francis Group
Boca Raton London New York

CRC Press is an imprint of the
Taylor & Francis Group, an **informa** business

CRC Press
Taylor & Francis Group
6000 Broken Sound Parkway NW, Suite 300
Boca Raton, FL 33487-2742

First issued in paperback 2020

ISBN 13: 978-0-367-57127-6 (pbk)
ISBN 13: 978-1-138-03316-0 (hbk)

Library of Congress Cataloging-in-Publication Data
Names: Lai, Xin Min, author. | Fu, Ming Wang, author. | Peng, Lin Fa, author. Title: Sheet metal meso- and microforming and their industrial applications/Xin Min Lai, Ming Wang Fu, Lin Fa Peng. Description: First edition. | Boca Raton, FL : CRC Press/Taylor & Francis Group, 2019. | "A CRC title, part of the Taylor & Francis imprint, a member of the Taylor & Francis Group, the academic division of T&F Informa plc." | Includes bibliographical references and index. Identifiers: LCCN 2018014812| ISBN 9781138033160 (hardback : acid-free paper) | ISBN 9781138033177 (ebook) Subjects: LCSH: Micromachining. | Sheet-metal work. Classification: LCC TJ1191.5 .L35 2019 | DDC 671.3/5--dc23 LC record available at https://lccn.loc.gov/2018014812

Visit the Taylor & Francis Web site at
http://www.taylorandfrancis.com

and the CRC Press Web site at
http://www.crcpress.com

Contents

Preface

The traditional sheet-metal forming has been widely used in many industrial clusters, which include automotive, aerospace, consumer electronics, biomedical, etc. This application is generally considered as macroscaled forming as the parts made by the manufacturing process usually have large size and big dimensions, such as beverage cans and cooking utensils. The knowledge and knowhow using this process for mass production, in terms of product design, process determination, tooling design and fabrication, and product quality assurance and control, are well established despite the fact that most of them are still experience-based knowhow and heuristic guidelines obtained and developed via extensive experimentation and trial and error. The developed knowledge and knowhow, however, play a critical role in metal-forming industries and facilitate the design and development of metal-formed products.

When the size of metal-formed parts is scaled down to meso- or microscale with at least two dimensions in the size range of a few millimeters or less than 1 mm, the metal forming belongs to the so-called meso- and microforming specified in this book. The data, information, and knowledge obtained or established in macroscaled forming may not be fully valid in meso- and microforming. In the past few decades, many efforts and research endeavors have been performed to study this unique forming technology in such a way to systematically establish a systematic knowledge system for the meso- and microforming of sheet metals.

Meso- and microforming of sheet metals refer to the forming process for fabrication of meso- and microscaled sheet metal parts and components via plastic deformation of sheet metals. In the past few decades, plenty of efforts from both academia and industry have been put forth to conduct extensive research on this unique forming process from different aspects including process development, tooling design, product quality assurance and control, in addition to exploration and investigation of some eluded and tantalized issues such as size effect and size effect affected deformation behavior, process performance, and the scatter of physical behaviors and phenomena in the forming process. These state-of-the-art research outcomes and findings need to be systematically summarized and archived, and thus this monograph has been written and published.

In this book, Chapter 1 first introduces the multiscale manufacturing processes using plastic deformation of materials. The two types of meso- and microscaled deformation processes, viz., meso- and microscaled bulk and sheet metal forming, are summarized and their uniqueness is presented. The fundamentals and difficulties in geometry and shape forming, quality and property assurance, and the promising applications and challenges of meso- and microforming are given. Chapter 2 presents the size effect of meso- and microforming. The size effect affected deformation behaviors and modeling of the deformation by different models and modeling approaches are delineated. Finally, the model applications and verifications are also presented.

In Chapter 3, experimental observations of geometry and grain size effects on friction behavior in meso- and microforming are articulated. How the geometry size effect affects friction behavior under liquid lubricating conditions in the tooling workpiece interface is extensively discussed. The observation of grain size effect on friction is also presented, and different frictional mechanisms are analyzed and discussed. In addition to frictional phenomena, the instability and formability of materials in meso- and microforming are also crucial. In Chapter 4, how size effect affects the forming limit of sheet metals under different geometry and grain size conditions is presented and elucidated. The modeling of fracture behaviors of sheet metals in meso- and microforming by different fracture models is presented. All these help to understand the instability and formability of materials in this small-scale metal-forming arena.

In Chapter 5, the detailed meso- and microforming processes for making sheet metal parts and components are presented. The processes including bending, deep drawing, hydroforming, and flexible forming of sheet metals by soft punch and the detailed process characteristics are delineated. The process realization is also depicted, and the quality of the formed parts is analyzed. All of these provide a basis for real applications of meso- and microforming of sheet metals. In Chapter 6, the main issues in meso- and microstamping of miniaturized sheet metal products for industrial applications are summarized. The applications in different clusters are articulated. By using the stamped metallic bipolar plate (BPP) as a typical case study, the complete design and fabrication of BPPs by the meso- and microforming process for their application in fuel cell is given and the quality-related issues are analyzed. The applications show the capability of the process and further promising applications.

Last but not least, a unique application of meso- and microforming of sheet metals for making meso- and microscaled bulk metal parts is presented in Chapter 7. The bulk forming of sheet metals to produce small bulk metal components via progressive and compound forming process is an advanced and innovative application of meso- and microforming. The process characteristics and uniqueness are delineated, and the detailed deformation behaviors and process performance are systematically described. The process shows an efficient and feasible approach for mass production of meso- and microscaled bulk metal parts using the progressive sheet metal-forming process.

Metal forming is one of the most efficient manufacturing processes. It has been widely used in making parts and components for many different industrial clusters due to its high productivity, good product quality, and low production cost. This conventional manufacturing process has been revitalizing for its new advancement and technology innovation. Meso- and microforming is one of them. This book summarizes the advancements and innovation of the process and it is state of the art in this arena for students, researchers, and engineers in industries, R&D organizations, and academia to learn and apply this advanced and practical manufacturing process.

Xin Min Lai
Ming Wang Fu
Lin Fa Peng

Acknowledgments

We are indebted to the following institutions and individuals who have helped to make this book possible due to their generous support and contributions to the research activities from which the research outcomes and findings summarized in this book arise.

- First, we would like to thank Professor Lin Zhongqin, president of Shanghai Jiaotong University and member of the Chinese Academy of Engineering, for his strong support, close supervision, and invaluable advice on the projects from which the knowledge and findings presented in this book arise.
- The collaboration between Shanghai Jiao Tong University (SJTU) and The Hong Kong Polytechnic University (HK PolyU) facilitated the research presented in this book to be successfully conducted, and both institutions provided various scholarships, grants, and research resources, which are pertinent to establishing research facilities, conducting collaboration research, and joint supervision of Ph.D. students. All of these contributed to the publication of this book.
- National Science Foundation of China for funding the projects with the project numbers 50820125506, 51235008, 51522506, 51575465, 51275294, 50805092, and 51575465, and the Shuguang Program funded by Shanghai Education Development Foundation and Shanghai Municipal Education Commission (17SG13), and the project of 152792/16E (B-Q55M) from the General Research Fund of Hong Kong Government support the researches summarized in this book.
- Dr. Zu Tian Xu, Dr. Yun Jun Deng, Dr. Dian kai Qiu, Dr. Peng Hu, Dr. Zhao Yang Gao, Mr. Meng Yun Mao, Mr. En Zhe Bao, Mr. Jin Wang, Dr. Bao Meng, and Dr. Ji Lai Wang for contributing part of their research results when they worked under the earlier projects.

Finally, we would like to express our sincere gratitude to our family members for supporting us throughout research and publishing process. Their understanding and support were crucial and invaluable for the completion of the earlier research projects and this book.

Xin Min Lai
Ming Wang Fu
Lin Fa Peng

Authors

Xin Min Lai is the Changjiang Scholar Chair Professor in the School of Mechanical Engineering, Shanghai Jiao Tong University. He received his B.Eng. degree in Mechanical Engineering from the Hebei Institute of Technology and M.Eng. and Ph.D. degrees in Mechanical Engineering from Tianjin University in China. He worked at the Hebei Institute of Technology as a faculty member for 12 years. In 1999, he joined SJTU as a faculty member and then promoted to full professor in 2001. From then on, he has been engaging in the research on meso-/micromanufacturing, new energy device development, assembly modeling of sheet-metal product, digital manufacturing of autobody, etc. Currently, he is the director of Shanghai Key Laboratory of Digital Manufacture for Thin-Walled Structures, a member of autobody technology committee of Society Automotive Engineering (SAE), China. He has been granted with five national and provincial awards, including the National Award for Science and Technology Progress and the Shanghai Award for Science and Technology Invention, etc. In addition, he also successively obtained the title of Scholar of Shanghai Shu Guang Training Program and the First Prize of General Motors Innovative Talent Award. He has already published more than 100 Science Citation Index (SCI) academic articles and has been granted more than 30 patents. He has secured a number of prestigious research projects which include 973 project, National Natural Sciences Foundation projects, and international joint research projects.

Recently Related Publications

1. Gao ZY, Peng LF, Yi PY, **Lai XM**. Grain and geometry size effects on plastic deformation in roll-to-plate micro/meso-imprinting process. *Journal of Materials Processing Technology*, 2015, 219: 28–41.
2. Xu ZT, Peng LF, Fu MW, **Lai XM**. Size effect affected formability of sheet metals in micro/meso scale plastic deformation: experiment and modeling. *International Journal of Plasticity*, 2015, 68: 34–54.
3. Mao MY, Peng LF, Yi PY, **Lai XM**. Modeling of the friction behavior in metal forming process considering material hardening and junction growth. *Journal of Tribology*, 2016, 138(1): 012202.
4. Yi PY, Wu H, Zhang CP, Peng LF, **Lai XM**. Roll-to-roll UV imprinting lithography for micro/nanostructures. *Journal of Vacuum Science & Technology B*, 2015, 33(6): 060801.

5. Zhang CP, Yi PY, Peng LF, **Lai XM**, Ni J. Fabrication of moth-eye nanostructure arrays using roll-to-roll UV-nanoimprint lithography with an anodic aluminum oxide mold. *Nanotechnology, IEEE Transactions on*, 2015, 14(6): 1127–1137.

6. Wang J, Yi PY, Deng YJ, Peng LF, **Lai XM**, Ni J. Mechanism of forming defects in roll-to-roll hot embossing of micro-pyramid arrays: II. Numerical study. *Journal of Micromechanics and Microengineering*, 2015, 25(11): 115030.

7. Yi PY, Shu YY, Deng YJ, Peng LF, **Lai XM**. Mechanism of forming defects in roll-to-roll hot embossing of micro-pyramid arrays I: experiments. *Journal of Micromechanics and Microengineering*, 2015, 25(10): 105017.

8. Deng YJ, Yi PY, Peng LF, **Lai XM**. Flow behavior of polymers during the roll-to-roll hot embossing process. *Journal of Micromechanics and Microengineering*, 2015, 25(6): 065004.

9. Jiang TH, Peng LF, Yi PY, **Lai XM**. Analysis of the electric and thermal effects on mechanical behavior of SS304 subjected to electrical-assisted forming process. *Journal of Manufacturing Science and Engineering*, 2015, 138(6): 061004.

10. Xu ZT, Peng LF, **Lai XM**. Electrically assisted solid-state pressure welding process of SS 316 sheet metals. *Journal of Materials Processing Technology*, 2014, 214(11): 2212–2219.

11. Xu ZT, Peng LF, **Lai XM**, Fu MW. Geometry and grain size effects on the forming limit of sheet metals in micro-scaled plastic deformation. *Materials Science and Engineering: A*, 2014, 611: 345–353.

12. Peng LF, Mai JM, Jiang TH, **Lai XM**, Lin ZQ. Experimental investigation of tensile properties of SS316L and fabrication of micro/mesochannel features by electrical-assisted embossing process. *Journal of Micro and Nano-Manufacturing*, 2014, 2(2): 021002.

13. Deng YJ, Yi PY, Peng LF, **Lai XM**, Lin ZQ. Experimental investigation on the large-area fabrication of micro-pyramid arrays by roll-to-roll hot embossing on PVC film. *Journal of Micromechanics and Microengineering*, 2014, 24(4): 045023.

14. Peng LF, Deng YJ, Yi PY, **Lai XM**. Micro hot embossing of thermoplastic polymers: a review. *Journal of Micromechanics and Microengineering*, 2014, 24(1): 013001.

15. Gao ZY, Peng LF, Yi PY, **Lai XM**. An experimental investigation on the fabrication of micro/meso surface features by metallic roll-to-plate imprinting process. *Journal of Micro and Nano-Manufacturing*, 2013, 1(3): 031004.

16. Mai JM, Peng LF, **Lai XM**, Lin ZQ. Electrical-assisted embossing process for fabrication of micro-channels on 316L stainless steel plate. *Journal of Materials Processing Technology*, 2013, 213(2): 314–321.

17. Peng LF, Yi P, **Lai XM**. Design and manufacturing of stainless steel bipolar plates for proton exchange membrane fuel cells. *International Journal of Hydrogen Energy*, 2014, 39(36): 21127–21153.

18. Qiu DK, Yi P, Peng LF, **Lai XM**. Channel dimensional error effect of stamped bipolar plates on the characteristics of gas diffusion layer contact pressure for proton exchange membrane fuel cell stacks. *Journal of Fuel Cell Science and Technology*, 2015, 12(4): 041002.

19. Qiu DK, Yi P, Peng LF, **Lai XM**. Study on shape error effect of metallic bipolar plate on the GDL contact pressure distribution in proton exchange membrane fuel cell. *International Journal of Hydrogen Energy*, 2013, 38(16): 6762–6772.

Ming Wang Fu received his B.Eng. and M.Eng. degrees in materials science and engineering from the Northwestern Polytechnic University in Xi'an, China, and Ph.D. degree in mechanical engineering from the National University of Singapore, Singapore. Before he went to Singapore for his career development, he had worked in China as a faculty member and conducted many projects funded by governmental agencies and industries. In 1991 and 1994, he received the honorary awards of Outstanding Young Teacher and Outstanding Teacher from the Ministry of Aeronautic and Astronautic Industries of the People's Republic of China. During his eight and a half year tenure of the faculty appointment in the Mainland China, he was promoted to associate and full professor via the fast track promotion scheme in 1992 and 1995, respectively. In 1997, he joined the Singapore Institute of Manufacturing Technology as a senior research engineer. In August 2006, he joined the HK PolyU as a faculty member. His research endeavors include integrated product and process design and development, metal-forming technologies, dies and molds CAD, microscaled product development and processing of advanced materials, including Ti-alloys and bulk metallic glasses. He is also on the editorial board of a number of longstanding *International Journals which include International Journal of Plasticity, Materials and Design, International Journal of Damage Mechanics, International Journal of Advanced Manufacturing Technology*, and the *Chinese Journal of Mechanical Engineering*. He is also serving as an invited reviewer for many SCI journals, prestigious project awards, and funding applications. He has over 170 papers published by SCI journals and four monographs arising from his researches.

Monograph

1. **MW Fu**. *Design and Development of Metal Forming Processes and Products aided by Finite Element Simulation*, Springer, London, 2016.
2. **MW Fu**, WL Chan. *Micro-Scaled Product Development via Microforming: Deformation Behaviors, Processes, Tooling and Its Realization*, Springer, London, 2014.
3. JY Fuh, YF Zhang, AYC Nee, **MW Fu**. *Computer-Aided Injection Mould Design and Manufacture*, Marcel Dekker, Inc., New York, 2004.

Recently Related Publications

1. B Meng, **MW Fu**, SQ Shi. Deformation behavior and microstructure evolution in thermal-aided microforming of titanium dental abutment. *Materials & Design*, 2016, 89: 1283–1293.

2. B Meng, **MW Fu**, CM Fu, KS Chen. Ductile fracture and deformation behaviors of pure copper in progressive microforming. *Materials & Design*, 2015, 83: 14–25.

3. B Meng, **MW Fu**. Size effect on deformation behavior and ductile fracture in microforming of pure copper sheets considering free surface roughening. *Materials & Design*, 2015, 83(2015): 400–412.

4. ZT Xu, LF Peng, **MW Fu**, XM Lai. Size effect affected formability of sheet metals in micro/meso scale plastic deformation: experiment and modeling. *International Journal of Plasticity*, 2015, 68: 34–54.

5. JQ Ran, **MW Fu**. A hybrid model for analysis of ductile fracture in micro-scaled plastic deformation of multiphase alloys. *International Journal of Plasticity*, 2014, 61: 1–16.

6. ZT Xu, LF Peng, XM Lai, **MW Fu**. Investigation of geometry and grain size effects on the forming limit of sheet metals in micro-scaled plastic deformation. *Materials Science and Engineering A*, 2014, 611: 345–353.

7. JQ Ran, **MW Fu**. Applicability of ductile fracture criteria in micro-scaled plastic deformation. *International Journal of Damage Mechanics*, 2014.

8. B Meng, **MW Fu**, CM Fu, JL Wang. Multivariable analysis of micro shearing process customized for progressive forming of micro-parts. *International Journal of Mechanical Sciences*, 2015, 93: 191–203.

9. HS Liu, **MW Fu**. Prediction and analysis of ductile fracture in sheet metal forming; Part I: a modified Ayada criterion. *International Journal of Damage Mechanics*, 2014.

10. HS Liu, **MW Fu**. Prediction and analysis of ductile fracture in sheet metal forming: Part II: application of a modified ductile criterion. *International Journal of Damage Mechanics*, 2014.

11. JL Wang, **MW Fu**, JQ Ran. Analysis of size effect on flow-induced defect in micro scaled forming process. *International Journal of Advanced Manufacturing Technology*, 2014, 16: 421–432.

12. JL Wang, **MW Fu**, JQ Ran. Analysis of the size effect on springback behavior in micro scaled U-bending process of sheet metals. *Advanced Engineering Materials*, 2014, 16(4): 421–432.

13. JL Wang, **MW Fu**, JQ Ran. Analysis and avoidance of flow-induced defect in meso-scaled plastic deformation via simulation and experiment. *International Journal of Advanced Manufacturing Technology*, 2013, 68: 1551–1564.

14. **MW Fu**, WL Chan. Micro-scaled progressive forming of bulk micropart via directly using sheet metals. *Materials & Design*, 2013, 49: 774–783.

15. JQ Ran, **MW Fu**, WL Chan. The influence of size effect on the ductile fracture in micro-scaled plastic deformation. *International Journal of Plasticity*, 2013, 41: 65–81.

16. **MW Fu**, WL Chan. A review of the state-of-the-art microforming technologies. *International Journal of Advanced Manufacturing Technology*, 2013, 67: 2411–2437.

17. WL Chan, **MW Fu**. Meso-scaled progressive forming of bulk cylindrical and flanged parts using sheet metal. *Materials & Design*, 2013, 43: 249–257.

18. **MW Fu**, B. Yang, WL Chan. Experimental and simulation studies of micro blanking and deep compound process using copper sheet. *Journal of Materials Processing Technology*, 2013, 213: 101–111.

Lin Fa Peng is a professor in the Mechanical School of Shanghai Jiao Tong University. He received his B.Eng. and M.Eng. degrees in Materials Science and Engineering from Jilin University in Changchun, China and Ph.D. degree in Mechanical Engineering from SJTU, China. In 2010, he joined the Mechanical School of SJTU as a faculty member after he completed his 2-year postdoctoral research work. His research specialties include material processing technology, micro-/mesoforming, process modeling and optimization. Currently, his research focus is on mechanics analysis of deformation processes at micro/mesoscale, material characterization of metal materials, and forming process design and its application. He has been funded by three projects from the National Science Foundation of China (NSFC) as PI, and participated in two key projects from NSFC. He has published more than 40 papers in SCI journals and has been granted more than 20 Chinese patents of invention. In addition, he received 2011 National Excellent Doctoral Dissertation Award Nomination and the title of 2012 Shanghai Rising-Star by Shanghai Municipal Government. In 2015, he was granted the National Natural Science Foundation—Outstanding Youth Foundation project and received the first prize of Natural Science Award by the Ministry of Education, the People's Republic of China.

Recently Related Publications

1. **LF Peng**, ZT Xu, ZY Gao, MW Fu. A constitutive model for metal plastic deformation at micro/meso scale with consideration of grain orientation and its evolution. *International Journal of Mechanical Sciences*, 2018, 138–139: 74–85.
2. ZT Xu, **LF Peng**, EZ Bao. Size effect affected springback in micro/meso scale bending process: experiments and numerical modeling. *Journal of Materials Processing Technology*, 2018, 252: 407–420.
3. **LF Peng**, ZT Xu, MW Fu, XM Lai. Forming limit of sheet metals in mesoscale plastic forming by using different failure criteria. *International Journal of Mechanical Sciences*, 2017, 120: 190–203.
4. MY Mao, **LF Peng**, P Yi, XM Lai. Modeling of the friction behavior in metal forming process considering material hardening and junction growth. *Journal of Tribology*, 2016, 138(1): 012202.
5. **LF Peng**, MY Mao, MW Fu, XM Lai. Effect of grain size on the adhesive and ploughing friction behaviours of polycrystalline metals in forming process. *International Journal of Mechanical Sciences*, 2016, 117: 197–209.
6. **LF Peng**, PY Yi, P Hu, XM Lai, J Ni. Analysis of micro/mesoscale sheet forming process by strain gradient plasticity and its characterization of tool feature size effects. *Journal of Micro and Nano-Manufacturing*, 2015, 3(1): 011006.
7. ZY Gao, **LF Peng**, PY Yi, et al. Grain and geometry size effects on plastic deformation in roll-to-plate micro/meso-imprinting process. *Journal of Materials Processing Technology*, 2015, 219: 28–41.

8. ZT Xu, **LF Peng**, MW Fu, et al. Size effect affected formability of sheet metals in micro/meso scale plastic deformation: experiment and modeling. *International Journal of Plasticity*, 2015, 68: 34–54.

9 ZT Xu, **LF Peng**, XM Lai, et al. Geometry and grain size effects on the forming limit of sheet metals in micro-scaled plastic deformation. *Materials Science and Engineering: A*, 2014, 611: 345–353.

10. **LF Peng**, JM Mai, T Jiang, et al. Experimental investigation of tensile properties of ss316l and fabrication of micro/mesochannel features by electrical-assisted embossing process. *Journal of Micro and Nano-Manufacturing*, 2014, 2(2): 021002.

1

Introduction

1.1 Introduction

Manufacturing is referred to as the process in which raw materials are transformed into its final product with designed geometries and shapes, and the required properties and qualities are improved using the specific tooling and equipment. The manufacturing process begins with the creation of materials from which the design is made, and the materials are then modified in this process to become products. In the manufacturing process, the mass or volume of materials can be increased, decreased, or made constant. Therefore, the manufacturing process can be classified into additive, subtractive, and constant-mass manufacturing. Additive manufacturing is defined as the process of joining materials to make objects, according to the three-dimensional CAD models, usually layer upon layer in which the mass and volume of materials are increased and is thus named additive. Additive manufacturing has different synonyms such as additive fabrication, additive layer manufacturing, layer manufacturing, etc. Opposed to additive manufacturing, subtractive manufacturing is referred to as the traditional machining process in which the raw material block is cut into the desired final shape and size by controlled material-removal process. In this process, achieving the shape, geometry, tolerance, and surface quality of the machined workpiece is critical. The constant-mass manufacturing, on the other hand, is defined as the process in which the materials are transported from one place to the other, and in such a way, the shapes and dimensions of the material block are changed. In this material deformation process, the mass of the material remains the same, and thus the volume of material is generally also constant. Deformation-based manufacturing belongs to the category of manufacturing process that includes bulk forming and sheet metal-forming processes.

In the manufacture of parts and components with different sizes and geometries, the size scale of the fabricated workpiece needs to be considered in process route design and process parameter determination. Traditionally, the manufactured workpieces are quite large with the size length in macroscale. With the ubiquitous trend of product miniaturization in many industrial clusters, the size scale of the fabricated workpiece is downscaled to the length of a few millimeters and submillimeters, and therefore, the so-called meso- and microscale manufacturing emerges as a popular and over-whelming manufacturing process. In deformation-based manufacturing via plastic deformation of materials for making parts and components, the definition of size scale is based on the dimensions of manufactured work-pieces, which could be different from those in other fields and disciplines. In the metal-forming arena, microscale is referred to as the size range of work-pieces with at least two dimensions in submillimeters, while the mesoscale is defined as the size scope from 1 to 10 mm, and similarly, the parts should have at least two dimensions in this size range [1]. The traditional macroscale is thus defined as the size category of workpieces larger than 10 mm [1]. The miniaturized parts are generally referred to as meso- and microscaled work-pieces. The deformation-based manufacturing processes employed to fabri-cate those miniaturized parts are termed as meso- and micromanufacturing processes.

Figure 1.1 shows the workpieces made by deformation-based manufac-turing, viz., metal-forming processes. The parts can be classified into the

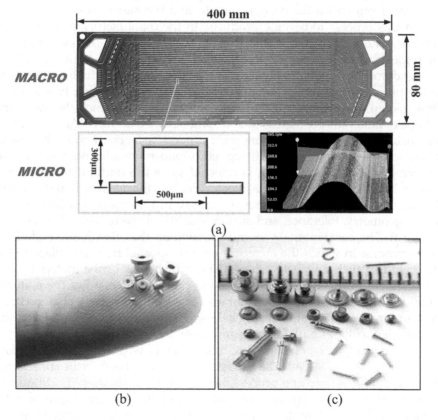

FIGURE 1.1
The metal-formed workpieces with different size scales.

following size categories, viz., macro-, meso-, and microscale. Figure 1.1a shows the macroscaled workpiece as its size scale falls within the macroscale definition with dimensions of 400×80 mm. Figure 1.1b and c present the meso- and microscaled parts, as their two dimensions are greater and less than 1.0 mm, respectively. The macroscaled part, however, has many microfeatures, and the formation of these miniaturized features needs to employ meso- and microforming processes. Therefore, manufacturing of the macroscaled part with various microfeatures also belongs to meso- and microforming category.

1.2 Meso- and Microscaled Bulk Metal Forming

Metal forming is an efficient manufacturing process for making net shape or near-net shape parts and components by employing the plasticity of materials to deform the materials. This traditional metal-forming process was a common practice for making simple tools a few thousand years ago. The blacksmith in ancient times used tools such as a hammer and an anvil to deform heated metals. Nowadays, this process offers many unique and attractive advantages, such as high productivity, low production cost, excellent material utilization, superior mechanical properties, and complex geometries of the deformed parts and components using the modern forming equipment [2]. With the ever-increasing costs of materials, energy, and manpower, and increasingly strict statutory regulations arising from environment-friendly and sustainable development, this conventional manufacturing technology is now facing more new challenges in terms of productivity, cost, and quality.

In metal forming, the plastic deformation of raw materials in the form of blank or billet with different geometries and shapes, including plate, sheet, bar, rod, block, wire, tube, etc., is conducted, and the material block or sheet is deformed into desirable geometries without the change of mass and composition of the materials [2–5]. The deformation of materials can be classified into two types of forming processes, viz., bulk and sheet metal forming. The bulk forming refers to the deformation of three-dimensional material block into the deformed parts and components with a specific volume. Bulk metal forming includes forging, extrusion, rolling, drawing, etc., which is used to prepare material preforms and net shape or near-net shape parts and components. Sheet metal forming refers to the metal working process to deform or stamp sheet metals into sheet metal parts and components with specific geometries, dimensions, and tolerance and quality requirements. In sheet metal forming, the surface of the sheet plays an important role, and the interfacial interaction between the sheet metal and tooling surfaces affects process performance and the quality of deformed parts. The main forming operations include shearing, drawing, bending, and forming. In detail, each operation can have its detailed subset operations. Shearing can have more detailed shearing-based operations such as blanking, punching or piercing, notching, and trimming. Forming can also have its subset operations including stretch forming, flexible die forming, bulging, spinning, peen forming, superplastic forming, etc. [2].

In bulk forming, the sizes of metal-deformed parts can be classified into three scales, viz., macro-, meso-, and microscales (Figure 1.1). From the perspective of shape- and geometry-forming property and quality assurance, the challenges in metal forming include the following aspects: (a) high shape and geometry accuracy and small dimensional tolerance, (b) good surface quality, surface texture, and property, (c) desirable microstructure, phase composition, and distribution, (d) required local and global deformation flow line and texture, (e) high mechanical, thermal, and physical properties, (f) defect-free products, viz., without flow-induced defects induced by irregular material flow, stress-induced defects caused by deformation stress, and those generated by undesirable microstructure and its distribution. In hot, warm, and cold forming, these challenges could be different. When the geometrical size of the deformed parts is decreased to meso- and microscales, the aforementioned challenges become more critical. This is generated by the size effect and its affected behaviors and phenomena in the formation of miniaturized parts [6,7].

In meso- and microscaled plastic deformation, size effect is a nontrivial phenomenon, which can induce many different behaviors and phenomena. Currently, the physics and scientific nature behind size effect are not clearly understood, and the fundamental knowledge of size effect and its physical presence have not been fully established. Phenomenologically, size effect can result in many phenomena, which are different from those in macroscale. From a deformation perspective, it induces different mechanical and physical phenomena in microscale, such as different flow stress, deformation load, flow behavior, friction phenomenon, damage, and fracture characteristics of materials. From a forming process aspect, it causes different process performance in terms of geometry accuracy, dimensional tolerance, variable mechanical and physical properties, and many other quality-related indicators. In addition, one of the eluded issues is the scatter of geometry accuracy, dimensional tolerance, property and quality of the deformed part, and process parameters. The scatter leads to uncertainty and difficulty in accurate determination of process route, configuration of process parameters, and assurance and control of the property and quality. Taking the dimension of the deformed part as an instance, the scatter can lead to a 25%–30% fluctuation of the measured dimension in the same batch production. Therefore, how to address this issue is critical and nontrivial, as the physical nature of size effect needs to be figured out and the solutions to reduce or eliminate uncertainty and undesirable phenomena could be possibly developed.

From a process realization perspective, fabrication of meso- and microscaled bulk parts has three critical issues. The first one is to prepare the billet and preform with the same size scale. Using the traditional micromachining process, it is not cost-effective and productive. The second issue is billet positioning in die cavity and the transportation among different operation stations. For positioning, the billet needs to be exactly placed inside the die cavity with good alignment. However, the billet is too small and could stick to the surface of the handling tool or the wall of cavity, and it is thus difficult to be placed at the right position. In addition, the transportation of billet or preform from one operation to the other is also a tough issue for the same reason as positioning. The third difficulty is the ejection of the deformed part from the die cavity. The strong sticky force between the workpiece and tooling and the elastic recovery of deformation of the deformed parts make the ejection of the part from the die cavity to be difficult. In tandem with this, a progressive formation of bulk parts and components using sheet metal stamping and forming was developed to solve the aforementioned problems [8–12]. This promising meso- and microscaled formation of miniaturized bulk metal parts directly using sheet metal is presented in detail in Chapter 7.

1.3 Meso- and Microscaled Sheet Metal Forming

The definitions of meso- and microscales in bulk metal forming are easily understood and can be conveniently used in the classification of the formation of bulk materials. In progressive sheet metal formation for making the miniaturized bulk metal parts and components, the classification of macro-, meso-, and microscales can be done based on the dimensions of the formed bulk metal parts according to the previously described size scale definitions [8–12]. For sheet metal forming, however, the main size parameters are the thickness, width, area of sheet metals, and the dimensions of microscaled features in the formed sheet metal parts. Defining the size scale based on the characteristic size parameters and classifying the sheet metal formation into different size categories have not yet been done. There is also no consensus on the size scale definition among research and industrial communities. For the plane-strain deformation of sheet metals, the sheet thickness can be considered a key parameter for specifying the size scale of sheet metal forming, but there is no available literature reporting the definition of size scale using this parameter, and this should be easily addressed in future.

In sheet metal forming, local microfeatures often appear in macroscaled parts. Figure 1.1a shows a good instance with many microfeatures to be produced by forming process. For the size scale of these microfeatures, it can be classified based on the size scale definition of bulk metal forming. However, the deformation belongs to sheet metal forming as it has the deformation characteristics of meso- and microscaled sheet metal forming. The critical difference of meso- and microscaled sheet metal forming from bulk metal forming is that the thickness dimension of sheet metal forming is more miniaturized compared with the other two dimensions. Therefore, the meso- and microscaled sheet metal forming has one or two in-plane dimensions within the meso- and microscale [13–15]. In addition, their long microchannel feature is a unique characteristic [16–18].

In meso- and microscaled sheet metal forming, the relative material flow inside the sheet metal is small, but the rigid displacement of materials is large. This is caused by the difficulty of deformation in the thin deformation zone across the thickness direction and the two interfacial frictions on the contact surfaces of two sides. In addition, size effect affects deformation, leading to different deformation behaviors and process performances. By using the example shown in Figure 1.1a, the main issues closely related to the forming of microfeatures include (a) dimensional inaccuracy caused by uneven springback [19]; (b) nonuniform deformation induced by the interaction of intrinsic and geometric dimensions [20,21]; (c) formability and forming limit of the materials [22]; and (d) friction and contact condition variation [23]. The detailed discussion addressing these issues as well as the forming process of macroscaled sheet metal parts with many microscaled features will be extensively discussed in Chapter 6.

Although both meso- and microscaled bulk and sheet metal forming for making multiscaled sheet metal parts are introduced in this chapter, the focus of this book is more on meso- and microscaled sheet metal forming. In addition, size effect and its induced phenomena, such as size effect affected friction, formability, and instability, will be discussed. The details of meso- and microforming processes of sheet metals will also be presented and discussed. The applications of this unique forming process will be summarized. By using meso- and microscaled progressive sheet metal forming, the fabrication of meso- and microscaled bulk parts and components will be presented.

1.4 Summary

In this chapter, an introduction to multiscale manufacturing processes via the plastic deformation of materials is presented. Two types of meso- and micro-scaled deformation processes, viz., meso- and microscaled bulk and sheet metal forming, are summarized, and their uniqueness is given. The difficulty in shape and geometry forming and quality and property securing in forming is also presented. The chapter serves as a brief introduction to the whole book and presents a panorama of the fundamentals, promising applications and challenges in meso- and microscaled sheet metal forming.

References

1. M.W. Fu, W.L. Chan, *Micro-scaled Products Development via Microforming: Deformation Behaviours, Processes, Tooling and Its Realization*, Springer Science & Business Media, London, 2014.
2. M.W. Fu, *Design and Development of Metal-Forming Processes and Products Aided by Finite Element Simulation*, Springer, London, 2016.
3. T. Altan, G. Ngaile, G. Shen, *Cold and Hot Forging: Fundamentals and Applications*, ASM International, Materials Park, OH, 2005.
4. S. Kobayashi, S. Kobayashi, S.-I. Oh, T. Altan, *Metal Forming and the Finite-Element Method*, Oxford University Press on Demand, Oxford, OH, 1989.
5. V. Boljanovic, *Sheet Metal Forming Processes and Die Design*, Industrial Press Inc., New York, 2004.
6. M.W. Fu, J.L. Wang, A. Korsunsky, A review of geometrical and microstructural size effects in micro-scale deformation processing of metallic alloy components, *International Journal of Machine Tools and Manufacture* 109 (2016) 94–125.
7. M.W. Fu, W.L. Chan, A review on the state-of-the-art microforming technologies, *The International Journal of Advanced Manufacturing Technology* 67(9–12) (2013) 2411–2437.
8. W.L. Chan, M.W. Fu, Meso-scaled progressive forming of bulk cylindrical and flanged parts using sheet metal, *Materials & Design* 43 (2013) 249–257.
9. B. Meng, M.W. Fu, S.Q. Shi, Deformation behavior and microstructure evolution in thermal-aided mesoforming of titanium dental abutment, *Materials & Design* 89 (2016) 1283–1293.
10. B. Meng, M.W. Fu, C. Fu, K. Chen, Ductile fracture and deformation behavior in progressive microforming, *Materials & Design* 83 (2015) 14–25.
11. B. Meng, M.W. Fu, Size effect on deformation behavior and ductile fracture in microforming of pure copper sheets considering free surface roughening, *Materials & Design* 83 (2015) 400–412.
12. B. Meng, M.W. Fu, S.Q. Shi, Deformation characteristic and geometrical size effect in continuous manufacturing of cylindrical and variable-thickness flanged microparts, *Journal of Materials Processing Technology* 252 (2018) 546–558.
13. L.F. Peng, F. Liu, J. Ni, X.M. Lai, Size effects in thin sheet metal forming and its elastic–plastic constitutive model, *Materials & Design* 28(5) (2007) 1731–1736.
14. X.M. Lai, L.F. Peng, P. Hu, S.H. Lan, J. Ni, Material behavior modelling in micro/meso-scale forming process with considering size/scale effects, *Computational Materials Science* 43(4) (2008) 1003–1009.
15. L.F. Peng, P. Hu, X.M. Lai, D.Q. Mei, J. Ni, Investigation of micro/meso sheet soft punch stamping process – simulation and experiments, *Materials & Design* 30(3) (2009) 783–790.

16. L.F. Peng, X.M. Lai, D.A. Liu, P. Hu, J. Ni, Flow channel shape optimum design for hydroformed metal bipolar plate in PEM fuel cell, *Journal of Power Sources* 178(1) (2008) 223–230.
17. L.F. Peng, X.M. Lai, H.-J. Lee, J.-H. Song, J. Ni, Analysis of micro/mesoscale sheet forming process with uniform size dependent material constitutive model, *Materials Science and Engineering: A* 526(1) (2009) 93–99.
18. L.F. Peng, D.A. Liu, P. Hu, X.M. Lai, J. Ni, Fabrication of metallic bipolar plates for proton exchange membrane fuel cell by flexible forming process-numerical simulations and experiments, *Journal of Fuel Cell Science and Technology* 7(3) (2010) 031009–031009-9.
19. Z.T. Xu, L.F. Peng, E.Z. Bao, Size effect affected springback in micro/meso scale bending process: Experiments and numerical modeling, *Journal of Materials Processing Technology* 252 (2018) 407–420.
20. L.F. Peng, Z.T. Xu, Z.Y. Gao, M.W. Fu, A constitutive model for metal plastic deformation at micro/meso scale with consideration of grain orientation and its evolution, *International Journal of Mechanical Sciences* 138–139 (2017) 74–85.
21. M.W. Fu, W.L. Chan, Geometry and grain size effects on the fracture behavior of sheet metal in micro-scale plastic deformation, *Materials & Design* 32(10) (2011) 4738–4746.
22. L.F. Peng, Z.T. Xu, M.W. Fu, X.M. Lai, Forming limit of sheet metals in meso-scale plastic forming by using different failure criteria, *International Journal of Mechanical Sciences* 120 (2017) 190–203.
23. L.F. Peng, M.Y. Mao, M.W. Fu, X.M. Lai, Effect of grain size on the adhesive and ploughing friction behaviours of polycrystalline metals in forming process, *International Journal of Mechanical Sciences* 117 (2016) 197–209.

2

Size Effects in Meso- and Microscaled Forming

2.1 Introduction

New challenges emerge when the traditional macroscaled forming processes are extended to meso- and microscaled scenarios. Instead of a simple scale down of the traditional forming processes by a scaling factor, the meso- and microscaled forming processes, represented as meso- and microforming, behave distinctly different from the well-known macroscaled forming processes. The differences not only exist in one or more aspects but also in the many behaviors and phenomena in the entire small-scale deformation process, such as material strength, plastic behavior, springback, friction behavior, surface finishing, failure behavior, etc., which could be different to a certain extent or large degree from those in large scale. These differences are generally believed to be caused by the size change of the geometry and microstructural grain, or the variation of other physical properties or characteristics such as material density and surface features, and thus they are collectively referred as the size effect in meso- and microforming processes. Driven by the advantages of microforming, including high efficiency, low cost, good mechanical properties, and net shape in mass production of microparts, this unique process has thus emerged as one of the promising micromanufacturing processes. Numerous research efforts have therefore been carried out to explore the size effect and its affected deformation behavior, process performance, and microformed product quality with the aim of figuring out (a) whether the data, information, and knowledge obtained and established in the traditional forming arena are valid in meso- and microforming and (b) if not, how to leverage this information and knowledge into the meso- and microscaled one, which definitely cannot be just straightforwardly scaled down by a simple scaling factor.

In meso- and microforming, one of the main intrinsic reasons of size effect is the interaction of the scaled-down geometry and the scale-independent factors of materials such as microstructure, surface roughness, asperity, etc.

In the traditional forming process, the scale-independent factors are much smaller than the dimensions of the materials to be deformed. Nevertheless, these factors remain unchanged when the deformation body size is scaled down from macro- to meso- and microlevels. As a result, the overall behaviors of the deforming materials are affected and deviated from the descriptions articulated by traditional theory and knowledge. To characterize and describe the mechanism of size effect and its influence on different deforming behaviors, many experimental and modeling studies have been conducted. In this chapter, the size effect in plastic deformation of materials in meso- and microscales is discussed. Its affected deformation behaviors, especially for those in sheet metal forming including springback, forming limit, and ductile fracture, will be extensively discussed. The modeling studies of size effect based phenomena are summarized and how they affect the process performance is also articulated.

2.2 Scaling Law in Multiscaled Forming Processes

Scaling is an accepted strategy to study the multiscaled forming processes and the effect of miniaturization. For an easy evaluation of the scaled deformation, it is generally assumed that the nature and state of stress, strain, their rates, and the deformation mode remain constants in the scaling process. Therefore, the theory of similarity, especially for conforming strictly to the geometrical similarity, is complied with and well-suited for studies of multiscaled forming processes [1].

Based on the theory of similarity, the investigation of size effect in microforming can be conducted by employing the similarity between the traditional macroforming system and the microforming one, as well as the similarity of material properties and process conditions of these systems. In each forming system, there are geometrical and nongeometrical factors, which affect the performance of the forming systems. The geometrical factors principally refer to the geometrical-related parameters such as feature sizes of metal sheets, tool geometric dimensions, etc., in the system, while the nongeometrical factors include property-related parameters of materials, process-related factors such as deformation load, formability, frictional coefficient, etc.

For an ordinary forming system A, it is supposed that there are m geometrical factors and n nongeometrical factors. The geometrical parameters and nongeometrical parameters are described as the sets $M = \{S_1, S_2,...,S_m\}$ and $N = \{T_1, T_2,...,T_m\}$, respectively. Similarly, for a meso- or microforming process $\bar{A}$, it could be written as $\bar{M} = \{\bar{S}_1, \bar{S}_2,...,\bar{S}_m\}$ and $\bar{N} = \{\bar{T}_1, \bar{T}_2,...,\bar{T}_m\}$. According to the scaling law, if two manufacturing systems are similar, the following equation should be held [2]:

$$\frac{\overline{S_1}}{S_1} = \frac{\overline{S_2}}{S_2} = \cdots = \frac{\overline{S_m}}{S_m} = r_L \tag{2.1}$$

where r_L is the geometrical scaling factor.

Given a similar degree of nongeometrical similar elements of the similar microforming systems as

$$r_{Ti} = \frac{\overline{T_i}}{T_i} \tag{2.2}$$

The similarity of two systems can thus be evaluated by the following:

$$\psi_i = \left| r_{Ti} - r_L^k \right| \tag{2.3}$$

$$\Phi_i = \frac{\left| r_{Ti} - r_L^k \right|}{r_L} \tag{2.4}$$

where ψ_i is the similarity difference, Φ_i is the similarity accuracy, and k is the power parameter of the geometric dimension similarity. For nongeometrical properties, such as yield stress, flow stress, and elastic modulus, k does not change for different geometric dimensions and is equal to 0. For geometrical properties, the parameter k is different, e.g., it is equal to 1 for tool velocity and 2 for punch force and blank holder force in sheet-forming process [2].

According to the similar nature of similarity theory, similarity difference and similarity accuracy are generally used as benchmarks for comparing the microforming systems. If both the similarity difference ψ_i and the similarity accuracy Φ_i for forming systems A and B are equal to 0, it means that they are similar and no size effect has been observed. Otherwise, a size effect would exist in forming process.

2.3 Size Effect Affected Material Behaviors

In metal meso- and microforming, the deformation behavior of materials is influenced by the grain size, grain orientation, and feature size of materials. The most commonly used parameter to describe the deformation behavior of materials is the flow stress of materials, which determines the forming force, the load on the tools influencing the local flow behavior, and thus the filling of die cavities. In addition, it also further affects the quality and properties of microformed parts. With the increasing product miniaturization, multi-scaled deformations are widely used, and the size effect characterized by the variation of flow stress has attracted much more attention than ever before.

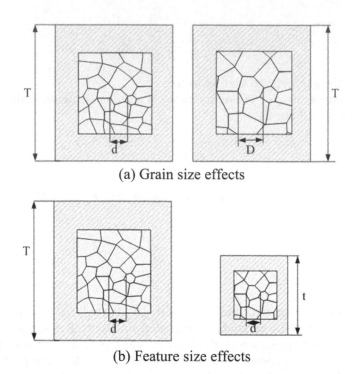

(a) Grain size effects

(b) Feature size effects

FIGURE 2.1
Grain and feature size effects with the decreasing of the scale [50].

Generally speaking, there are two different types of size effects that are widely considered to influence the material behavior. One is geometric size effect such as the variation of feature and specimen sizes, while the other is the microstructural size of materials, generally represented by grain size.

Figure 2.1 shows these two types of size effects. The grain size effect occurs when the grain size is increased from d' to d while the feature and specimen sizes are kept constant. The feature and specimen size effects, on the other hand, come into existence when the feature and specimen sizes are decreased from D to D' while the grain sizes of materials are kept constant. Currently, both size effects are widely studied in meso- and microforming.

For grain size effect, on the one hand, Hall [3] and Armstrong et al. [4] reported that the tensile yield, fracture stress, and the flow stress decrease with the increase of grain size of materials. This fact can be explained by the grain boundary strengthening caused by the pileup of dislocations. The ratio of the total grain boundary surface area to the material volume is decreased with the increase in grain size. This leads to the decrease of grain boundary strengthening effect and flow stress.

For the feature and specimen size effect, on the other hand, the flow stress of metallic materials with the constant grain size is generally decreased in the size scale of deformation-based manufacturing with the scaling down

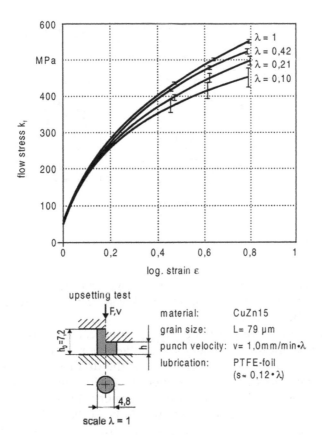

FIGURE 2.2
Flow stress versus logarithmic strain in the scaled upsetting tests [1].

of specimen dimensions. In the microupsetting experiments done by Geiger et al. [5], a series of specimen CuZn15 (grain size of 79 µm) have been made with different sizes and the same length scale λ to the dimension of tools. It is found that with the decrease of the length scale λ, the flow stress is also decreased, as shown in Figure 2.2.

Through microtensile test, Kals and Eckstein [6] used CuNi18Zn20 sheets (grain size of approximately 40 µm) of different scales to investigate the size effects of sheet material behavior. Figure 2.3 shows the flow stress curves for different length scales. It is also found that the flow stress increases with the length scale. In addition, with the same grain size and the decrease in specimen thickness, it is found that the ultrafine-grained Cu [7] has a slight stress change when a thicker specimen is used and the strain is 1%, but the stress drops obviously when the specimen thickness is reduced below 142 µm. For aluminum sheet, Raulea et al. [8,9] conducted uniaxial tensile and bending tests in which the grain size was kept constant, but the sample thickness changed. It is shown that the yield stress decreases with the number of

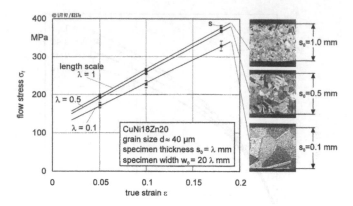

FIGURE 2.3
Flow stress of CuNi18Zn20 with different values of the length scale λ [6].

grains over the thickness up to one grain over the thickness. In addition, when the grain size is larger than the thickness, the yield strength tends to increase with the grain size.

To simultaneously consider both the geometrical and grain size effects, their interactive effect on the deformation behaviors of materials is quantified by the ratio of the specimen size (D is the diameter of bulk specimen and t is the thickness of sheet specimen) to grain size (d). Chan et al. [10] studied the stress–strain curves of pure copper with various ratios of the specimen size to grain size of materials. Both the flow stress and fracture strain decrease with the ratio. The research from Miyazaki et al. [11] also illustrated that the flow stress of polycrystalline Al, Cu, Cu-13 at% Al, and Fe decreases with a decrease in specimen thickness when the value of t/d becomes smaller than a fixed value, i.e., the critical value, independent of the amount of strain. In addition, different materials, such as pure nickel [12], aluminum, and brass [13], were further used in the experiments to study the effect of the ratio (t/d), and the same trend of the decrease of flow stress with the ratio was observed.

2.4 Flow Stress Modeling in Meso- and Microforming

From the perspective of design, flow stress and its hardening are very important factors, which affect material deformation, process design, and equipment selection. The flow stress in meso- and microscaled experiments was observed to be different from that of the conventional scale because of size effect, which needs to be more accurately modeled. In tandem with this, many constitutive models have been proposed to describe the size-dependent flow stress of metallic materials in microforming process. Among these models, surface layer model and grain boundary model are most representative and are summarized in the following.

For single crystal materials, according to crystal plastic theory and Schmid law [14], the critical shear resolved stress can be expressed as

$$\tau_R = (\cos\phi\cos\lambda)\sigma = \beta\sigma \quad \left(0 \le \beta \le \frac{1}{2}\right) \tag{2.5}$$

where β is the Schmid factor related to the grain orientation; ϕ is the angle between the normal stress σ and the normal direction of the slip plane; and λ is the angle between the slip direction and the normal stress. Therefore, the single crystal model can be designated as

$$\sigma_{sig}(\varepsilon) = m\tau_R(\varepsilon) \tag{2.6}$$

where m is the orientation factor ($m \ge 2$).

For polycrystalline materials at a relatively low temperature, one of the most widely accepted empirical theory relating the yield stress and the grain size is the Hall–Petch equation, which was further extended by Armstrong to include the flow stress and denoted as follows [15]:

$$\sigma(\varepsilon) = \sigma_0(\varepsilon) + \frac{k(\varepsilon)}{\sqrt{d}} \tag{2.7}$$

where d is the average grain size, $\sigma_0(\varepsilon)$ and $k(\varepsilon)$ are experimental constants at a specific strain ε. The first term $\sigma_0(\varepsilon)$ is known as the friction stress required to move individual dislocations in microyielded slip band pileups confined to isolated grains, whereas $k(\varepsilon)$ in the second term is the locally intensified stress needed to propagate general yield across the polycrystal grain boundaries [16]. Figure 2.4 shows the comparison between experimental and theoretical strength of mild steel varying with the grain size, derived from Armstrong's research [17]. It can be clearly seen that the relations of the low yield point and brittle fracture stresses of mild steel measured by Hall [3] with the grain size follow Eq. (2.7). Furthermore, Eq. (2.7) can also be applied for articulating the complete stress–strain behavior of the same material obtained by Armstrong et al. [4].

The value of $\sigma_0(\varepsilon)$ from Hall–Petch model (Eq. 2.7) is also related with the critical-resolved shear stress τ_R of a single crystal as

$$\sigma_o(\varepsilon) = M\tau_R(\varepsilon) \tag{2.8}$$

where M is the orientation factor related to the slips on deformation systems. According to Taylor model, an upper bound model, at least five active slip systems are required, whereas the Sachs model, a lower bound model, requires only one active slip system. For face centered-cubic (FCC) crystals,

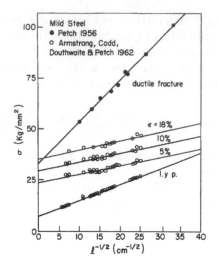

FIGURE 2.4
Comparison between the experimental and theoretical strengths of mild steel varying with the grain size [17].

M is equal 3.06 and 2.23 for the Taylor and Sachs modeling, respectively [16,18].

The polycrystal flow stress with the consideration of the grain size effect is thus obtained [4,19] as follows:

$$\sigma_{poly}(\varepsilon) = M\tau_R(\varepsilon) + \frac{k(\varepsilon)}{\sqrt{d}} \tag{2.9}$$

On the other hand, for the feature size effect, the decreasing flow stress with the increasing miniaturization has also been widely studied for explanation and articulation of size effects based on different assumptions. In the following, four different models are summarized.

2.4.1 Surface Layer Model

Surface layer model assumes that the workpiece to be deformed consists of the surface layer and the inner part. As shown in Figure 2.5, the surface layer has the thickness of one grain diameter and the remaining grains are the inner part of the workpiece. Compared with the inner grains, the surface layer grains are located at the free surface and thus have less restriction. It is harder for dislocations to pile up at grain boundaries when moving through the surface grains. Therefore, surface grains have less hardening and low resistance against deformation. In meso- and micro-forming, with the increase in miniaturization, the increase of the share of surface grains thus results in the decrease of flow stress of materials.

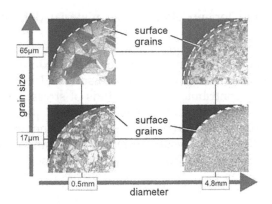

FIGURE 2.5
Surface layer model of size effect [1].

Therefore, the size effect thus occurs as the geometry size affects the material deformation behaviors.

2.4.1.1 Modeling

When the grain size of material is kept constant, the distribution of grains in the material cross-section changes with the decrease of feature size. As shown in Figure 2.6, with the decrease in size scale, the ratio of surface grains to inner grains increases and the material deformation behavior changes from polycrystal mode in macroscale to single crystal mode in microscale in the extreme scenario. For the meso- and microscale, which falls between the single crystal and polycrystalline ones, the following flow stress is thus satisfied:

$$\sigma_{sig} \leq \sigma_{micro/meso} \leq \sigma_{poly} \tag{2.10}$$

where σ_{sig} is the flow stress of single crystal, σ_{poly} is the flow stress of poly-crystal, and $\sigma_{micro/meso}$ is the flow stress of materials in meso- and microscale. It means that the single crystal model is the lower bound and the polycrystal

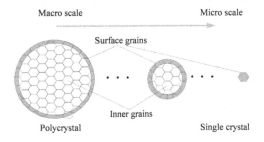

FIGURE 2.6
Grain distribution in a material section with the decreasing feature scale [50].

is the upper bound for modeling of the material deformation in these small scales.

Therefore, both the surface grains and inner grains play an important role in the mechanical response of materials to external loading under microscaled deformation. According to the surface layer model, the flow stress of material is contributed by two kinds of stresses, viz., the stresses of surface grains and inner grains. The mechanical properties of surface grains are rather similar to a single crystal and the inner grains to polycrystal. According to the single crystal model [Eq. (2.6)] and polycrystal model [Eq. (2.9)], the stress of surface grains, σ_s, and the stress of surface grains, σ_i, are thus represented as

$$\begin{cases} \sigma_s(\varepsilon) = m\tau_R(\varepsilon) \\ \sigma_i(\varepsilon) = M\tau_R(\varepsilon) + \dfrac{k(\varepsilon)}{\sqrt{d}} \end{cases} \tag{2.11}$$

Therefore, the flow stress of the surface layer model takes the following form:

$$\sigma(\varepsilon) = \frac{N_s\sigma_s(\varepsilon) + N_i\sigma_i(\varepsilon)}{N} = \frac{N_s m\tau_R(\varepsilon) + N_i\left(M\tau_R(\varepsilon) + k(\varepsilon)/\sqrt{d}\right)}{N} \tag{2.12}$$

where σ is the flow stress of material, N is the total number of grains, and N_s and N_i are the numbers of the surface and inner grains, respectively.

Considering $N_s = \eta N$, Eq. (2.12) can be expressed by using η, which is size factor, a size-dependent parameter used to evaluate the influence of size effect.

$$\sigma(\varepsilon, \eta) = \eta m\tau_R(\varepsilon) + (1 - \eta)\left(M\tau_R(\varepsilon) + \frac{k(\varepsilon)}{\sqrt{d}}\right) \tag{2.13}$$

When the size factor $\eta = 0$, the flow stress is converged to the polycrystalline model. While size factor $\eta = 1$, the flow stress follows the single crystal model.

From Eq. (2.13), the surface layer model can further be divided into two different terms: size-dependent model and size-independent model, as given by

$$\begin{cases} \sigma(\varepsilon) = \sigma_{ind} + \sigma_{dep} \\ \sigma_{ind} = M\tau_R(\varepsilon) + \dfrac{k(\varepsilon)}{\sqrt{d}} \\ \sigma_{dep} = \eta\left(m\tau_R(\varepsilon) - M\tau_R(\varepsilon) - \dfrac{k(\varepsilon)}{\sqrt{d}}\right) \end{cases} \tag{2.14}$$

In this way, the surface layer model is expressed as the conventional polycrystal material model by subtracting the influence of the size effect on flow stress.

For the size factor in surface layer model, there are two scenarios to be considered. One is bulk forming and the other is sheet metal forming. They are discussed in the following.

2.4.1.1.1 Size Factor in Micro Bulk Forming

For the size factor in micro bulk forming, Figure 2.7 shows the grain distribution in the section of a round specimen. It presents two kinds of grains: surface grains (Figure 2.7b) and inner grains (Figure 2.7c). The diameter of the specimen (feature size) is D and that of the grain size is d.

The number of surface grains can be calculated by Eq. (2.15) in the following:

$$\begin{cases} N_s = \dfrac{D^2\pi/4-(D-2d)^2\pi/4}{d^2\pi/4} = 4\left(\dfrac{D}{d}-1\right) & D \neq d \\ N_s = 1 & D = d \end{cases} \tag{2.15}$$

The total grain number in a section can be evaluated by

$$N = \frac{D^2\pi/4}{d^2\pi/4} = \left(\frac{D}{d}\right)^2 \tag{2.16}$$

Therefore, the size factor η, designated as the number ratio of surface grains to the total grains in the specimen, can be expressed as

$$\eta = \begin{cases} \dfrac{N_s}{N} = \dfrac{4d(D-d)}{D^2} & D \neq d \\ 1 & D = d \end{cases} \tag{2.17}$$

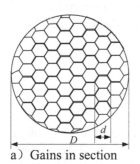
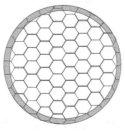
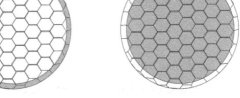

a) Gains in section   b) Outer/surface gains   c) Inner/volume gains

FIGURE 2.7
Surface and inner grains in the cross-section of a round specimen [50].

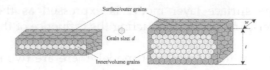

FIGURE 2.8
Surface and inner grains in the sheet metal specimen [50].

2.4.1.1.2 Size Factor in Micro Sheet Forming

For micro sheet forming, Figure 2.8 shows the grains of sheet metal in meso- and microforming. The sheet grain diameter is d. The sheet thickness and width are designated as t and w, respectively. The number of surface grains in the section can be calculated as follows.

$$\begin{cases} N_s = \dfrac{w+t}{d} - 2 \\ N_s = 1 \qquad\quad w = t = d \end{cases} \tag{2.18}$$

The number of total grains in a cross-section can be expressed as

$$N = \frac{wt}{d^2} \tag{2.19}$$

Therefore, the size factor is calculated as given by

$$\eta = \frac{N_s}{N} = \frac{d}{t} \cdot \frac{w+t-2d}{w} \tag{2.20}$$

Different from bulk forming, the size factor η in sheet metal meso- and microforming involves two parameters: the sheet width and thickness. When $w \to t$, the sheet can be considered as bulk, and the material behavior is similar to that of bulk forming as discussed earlier. In meso- and microforming, the sheet width w is usually much larger than the thickness t $(w \gg t)$, and the width w is also much larger than the grain diameter d $(w \gg d)$. Therefore, $(w + t - 2d)/w \approx 1$. Thus, Eq. (2.20) can be expressed as

$$\eta = \frac{d}{t} \tag{2.21}$$

2.4.1.2 Verification

Figure 2.2 shows the experimental result data from the research of Geiger et al., for micro bulk forming, and it can be used for verification of the surface layer model. The material used in the experiments is CuZn15 and the grain size is 79 μm. According to Eq. (2.17), the value of size factor η varies from 0.06 to 0.55 with different specimen diameters. Generally speaking, the flow stress curves can be represented by an exponential law. In addition, the least-square method can be used to calculate the undetermined coefficients in the model. Hence, the critical-resolved shear stress τ_R and $k(\varepsilon)/\sqrt{d}$ are obtained as follows:

$$\tau_R = 200\varepsilon^{0.43}, \quad \frac{k(\varepsilon)}{\sqrt{d}} = 80.41\varepsilon^{0.49}.$$

On the basis of Schmid law and single crystal plasticity theory, the flow stress for the single crystal model is $\sigma_{sig} = m\tau_R$ ($m \geq 2$). When the parameter m is 2, the flow stress σ_{sig} can be regarded as the lower bound condition of the surface layer model. The polycrystal model calculated according to Eq. (2.14) can act as the upper bound when the orientation factor M is determined as 3.06 according to the Taylor model. Figure 2.9 shows the comparison between the experimental and theoretical data based on the surface layer model. In addition, the surface layer model contains size-dependent and size-independent terms. The material model changes from single crystal model to the polycrystalline one when the size factor η varies from 0 to 1.0. Moreover, the simulated results calculated by the surface layer model match very well with the experimental results. Therefore, the model can be used to analyze the size effect and its affected deformation in meso- and microforming.

For sheet metal microforming, Peng et al. [20] conducted microtensile tests of SUS 304 stainless steel sheet. The sheet specimens with the average grain size of 25 μm were prepared with various sheet thicknesses (0.1, 0.2, 0.4, and 1.0 mm). Figure 2.10 shows the true strain and stress curves obtained from tensile experiments.

According to Eq. (2.21), the values of size factor η vary, respectively, from 0.025 to 0.25 for the different specimen thicknesses. For the scale factor $\eta = 0.025$ ($t = 1.0$ mm), the inner grains are much more than the surface grains; therefore, the flow stress can be regarded same as the conventional

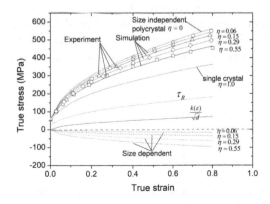

FIGURE 2.9
Comparison between the calculated and experimental results of bulk specimens [50].

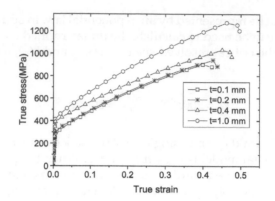

FIGURE 2.10
True strain and true stress of SUS304 sheets with various thicknesses [20].

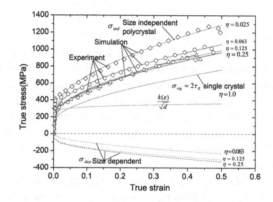

FIGURE 2.11
Comparison of the calculation and experimental results for microsheet forming [20].

polycrystalline material. Practically, the size effect is not considered in the traditional sheet-forming process by using sheets with a thickness of 1.00 mm. By using the similar method described in the earlier analysis of bulk microforming process, the critical-resolved shear stress τ_R and $k(\varepsilon)/\sqrt{d}$ are obtained as

$$\tau_R = 75 + 460\varepsilon^{0.59}, \quad \frac{k}{\sqrt{d}} = 200 + 186.2\varepsilon^{0.2}.$$

Figure 2.11 presents the simulation results based on the surface layer model, which match very well with the experimental ones in sheet metal microforming process.

In addition, Figure 2.12 shows the evolution of the size factor versus the sheet thickness for the sheets with different grain sizes. The size factor curves shift upward when the grain size is increased from 25 to 100 μm,

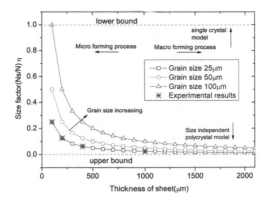

FIGURE 2.12
Size scale factor η as the function of grain size and sheet thickness [20].

which means that materials with larger grain sizes are prone to have size effects. Actually, the conventional macroforming and microforming can be defined according to the influence of size effect. The difference of the size scale can illustrate the reason why the flow stress of sheets with a thickness of 1.0 mm is much bigger than that of sheets with a thickness of 0.1 mm.

2.4.2 Grain Boundary Model

For the grain boundary model, as shown in Figure 2.13, it is assumed that the specimen is divided into surface layer and inner part, and the thickness of the surface layer is one grain, which is similar to the surface layer model. The grains in the inner part of the specimen are assumed to have two portions, viz., grain interior and grain boundary. The flow stress of these grains is determined by the "law of mixtures" based on the flow stresses of grain interior and grain boundary. For the surface layer of the specimen, it is assumed that no grain boundary is included in the grains since the grains located at the free surfaces have less hardening effect than the inner grains. In such a way, the grain boundary model is actually a composite model of grain and feature size effects of materials for meso- and microforming.

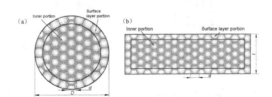

FIGURE 2.13
Surface layer and inner portions in a specimen section of (a) bulk specimen and (b) sheet specimen [24].

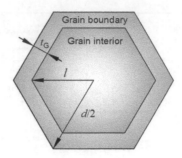

FIGURE 2.14
Grain interior and grain boundary models [24].

2.4.2.1 Modeling

Meyers and Ashworth [21] divided the grain into two portions, viz., grain interior and grain boundary, as shown in Figure 2.14. A composite model is thus proposed to calculate the flow stress of polycrystalline aggregate in the following.

$$\sigma_p = f_{GI}\sigma_{GI} + f_{GB}\sigma_{GB} \tag{2.22}$$

where σ_p is the flow stress of polycrystalline aggregate. f_{GI} and f_{GB} are the area fractions of grain interior and boundary, respectively. σ_{GI} and σ_{GB} are their corresponding flow stresses.

Assume that the grain size is d and the thickness of grain boundary layer is t_G. The size of grain interior l is thus $(d/2 - 2t_G/\sqrt{3})$. Hence, the area fractions of grain interior and boundary can be separately determined as:

$$f_{GI} = \frac{\dfrac{3\sqrt{3}}{2}\left(\dfrac{d}{2} - \dfrac{2}{\sqrt{3}}t_G\right)^2}{\dfrac{3\sqrt{3}}{2}\left(\dfrac{d}{2}\right)^2} = \left(1 - \dfrac{2}{\sqrt{3}} \cdot \dfrac{2t_G}{d}\right)^2 \tag{2.23}$$

$$f_{GB} = \frac{\dfrac{3\sqrt{3}}{2}\left(\dfrac{d}{2}\right)^2 - \dfrac{3\sqrt{3}}{2}\left(\dfrac{d}{2} - \dfrac{2}{\sqrt{3}}t_G\right)^2}{\dfrac{3\sqrt{3}}{2}\left(\dfrac{d}{2}\right)^2} = \dfrac{4}{\sqrt{3}} \cdot \dfrac{2t_G}{d}\left(1 - \dfrac{1}{\sqrt{3}} \cdot \dfrac{2t_G}{d}\right) \tag{2.24}$$

Substituting Eqs. (2.23) and (2.24) into Eq. (2.22), the flow stress for polycrystalline aggregate is designated as follows:

$$\sigma_p = \sigma_{GI} + \frac{8}{\sqrt{3}}t_G(\sigma_{GB} - \sigma_{GI})d^{-1} - \frac{16}{3}t_G^2(\sigma_{GB} - \sigma_{GI})d^{-2} \tag{2.25}$$

Considering the different area fractions of grain interior and grain boundary for different cross-sections, it is more reasonable to use the mean values of

t_G and d, i.e., $\bar{t}_G$ and $\bar{d}$, respectively [21]. Therefore, the following relationship is obtained:

$$\sigma_p = \sigma_{GI} + \frac{8}{\sqrt{3}}\bar{t}_G(\sigma_{GB} - \sigma_{GI})\bar{d}^{-1} - \frac{16}{3}\bar{t}_G^2(\sigma_{GB} - \sigma_{GI})\bar{d}^{-2} \qquad (2.26)$$

The relationships between $\bar{d}$ and d and $\bar{t}_G$ and t_G are further determined as [21]

$$\bar{d} = \frac{\pi}{4}d, \quad \bar{t}_G = 1.57t_G \qquad (2.27)$$

The term $\bar{t}_G\bar{d}^{-1}$ is approximately equal to $2t_G d^{-1}$. Furthermore, the thickness of grain boundary is generally related to the grain size in the following [21,22]

$$t_G = kd^n (0 < n < 1) \qquad (2.28)$$

where k and n are considered to be constant for a specific material.

Substituting Eqs. (2.27) and (2.28) into Eq. (2.26), the following relationship is thus established:

$$\sigma_p = \sigma_{GI} + \frac{16}{\sqrt{3}}k(\sigma_{GB} - \sigma_{GI})d^{n-1} - \frac{64}{3}k^2(\sigma_{GB} - \sigma_{GI})d^{2n-2} \qquad (2.29)$$

Equation (2.29) shows the effect of grain size on flow stress. With the decrease of grain size, the flow stress increases. When the grain size is larger (in micrometer range), $k^2 d^{2n-2}$ is much smaller than kd^{n-1} due to $0 < n < 1$, and the corresponding term can thus be eliminated. Equation (2.29) thus becomes

$$\sigma_p = \sigma_{GI} + \frac{16}{\sqrt{3}}k(\sigma_{GB} - \sigma_{GI})d^{n-1} \qquad (2.30)$$

When $n = 0.5$, Eq. (2.30) converges to the Hall–Petch relationship.

According to the surface layer model, the flow stress of material is represented by the weighted average of stresses of the inner part and the surface layer of specimen. The flow stress of inner grains can be expressed by that of polycrystalline aggregate, $\sigma_i = \sigma_p$, and the flow stress of surface layer is equal to that of grain interior, viz., $\sigma_s = \sigma_{GI}$. The geometry size effect is thus taken into account through the combination of surface layer model and composite model. The flow stress of material in microforming can be expressed as follows:

$$\sigma = \sigma_{GI} + \frac{16}{\sqrt{3}}(1-\eta)k(\sigma_{GB} - \sigma_{GI})d^{n-1} - \frac{64}{3}(1-\eta)k^2(\sigma_{GB} - \sigma_{GI})d^{2n-2} \quad (2.31)$$

According to the surface layer model in Section 2.4.1, for the bulk specimen, as shown in Figure 2.13a, the specimen diameter is D. η is calculated based on Eq. (2.17). For the sheet specimen with rectangular cross-section, as shown in

Figure 2.13b, the specimen thickness is t. η is thus determined by Eq. (2.21). Substituting Eqs. (2.17) and (2.21) into Eq. (2.31), the detailed constitutive models for bulk metal and sheet metal microforming can thus be obtained.

2.4.2.2 Verification

To verify the model, copper and copper alloys were used [22,23]. According to Eq. (2.29), the thickness of grain boundary is assumed to be

$$t_G = 0.133d^{0.7} \tag{2.32}$$

Therefore, the flow stress in Eq. (2.31) formulated for copper and copper alloys is given by

$$\sigma = \sigma_{GI} + \frac{16}{\sqrt{3}}(1-\eta) \times 0.133(\sigma_{GB} - \sigma_{GI})d^{-0.3} - \frac{64}{3}(1-\eta)$$
$$\times 0.133^2(\sigma_{GB} - \sigma_{GI})d^{-0.6} \tag{2.33}$$

The flow stresses of grain interior and grain boundary (σ_{GI} and σ_{GB}) are two unknowns. Therefore, more than two strain–stress curves of the deformation materials with different geometry size factors are needed to determine their values. Furthermore, Eq. (2.33) can be reformulated as a linear function of the geometry size effect factor.

$$\sigma = A + B(1-\eta) \tag{2.34}$$

where

$$\begin{cases} \sigma_{GI} = A \\ \dfrac{16}{\sqrt{3}} \times 0.133(\sigma_{GB} - \sigma_{GI})d^{-0.3} - \dfrac{64}{3} \times 0.133^2(\sigma_{GB} - \sigma_{GI})d^{-0.6} = B \end{cases} \tag{2.35}$$

Fitting the geometry size factor-flow stress curves with a linear function and solving Eq. (2.35), the values of σ_{GI} and σ_{GB} at a certain strain can be obtained. The flow stress curves of grain interior and grain boundary are finally determined. By using this method, Liu et al. [24] and Fu et al. [22] obtained the flow stress curves of grain interior and grain boundary of pure copper, as displayed in Figure 2.15. It is found that the flow stresses of grain boundary of pure copper from the two references are consistent with each other. And the mechanical behavior of grain boundary is illustrated to be harder than that of the grain interior.

The microtensile tests of pure copper sheets were conducted by Liu et al. [24]. Figure 2.16 shows the comparison of the calculated and experimented flow stresses for the same thickness sheet with different grain sizes, viz.,

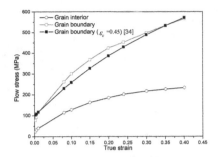

FIGURE 2.15
Flow stresses of grain interior and boundary of pure copper sheet [24].

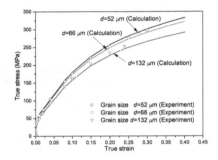

FIGURE 2.16
Influence of grain size effect on flow stress ($t = 0.4\,\text{mm}$) [24].

52, 66, and 132 μm. The flow stress is decreased with the increase of the average grain size. This phenomenon can be explained by Hall–Petch equation. In addition, a good agreement between the calculated and experimental result shows that the developed constitutive model, which considers both the geometry and grain size effects, can be used to model the deformation behavior in microforming process of sheet metal efficiently.

2.4.3 Crystal Plastic Model

Miyazaki et al. [11] found that the flow stress of polycrystalline Al, Cu, Cu-13 at% Al, and Fe decreases with the ratio of specimen thickness (t) to grain size (d) when the ratio (t/d) is smaller than a critical value. As shown in Figure 2.17, there are two common tendencies in all metals and alloys: (1) the critical value of t/d increases with the decrease in grain size; (2) extrapolation of curves takes the same flow stress at zero value of t/d independent of the grain size. Based on these experimental results, Miyazaki et al. [11] assumed that flow stress $\sigma(t)$ under a certain deformation can be represented by

$$\sigma(t) = \sigma_0 + \bar{H}(t)(\sigma_\infty - \sigma_0) \tag{2.36}$$

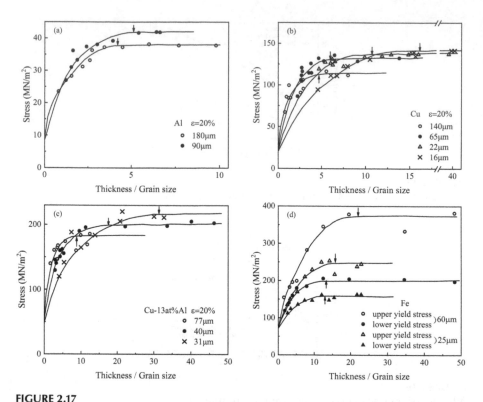

FIGURE 2.17
Thickness effect on the flow stresses in polycrystalline (a) Al, (b) Cu, (c) Cu-13 at%Al, and (d) Fe with various grain sizes [11].

where t is the thickness of the specimen, σ_0 is the value of the flow stress extrapolated to zero in specimen thickness, σ_∞ is a saturated value of the flow stress of thick specimens, and $\bar{H}(t)$ is the mean constraining force that stands for the grain interaction during deformation.

Based on the flow stress model in Eq. (2.36), a modified flow stress is proposed by Leu et al. [25] as

$$\sigma(t/d) = \sigma_0 + \alpha(t/d)(\sigma_\infty - \sigma_0) \tag{2.37}$$

where the dimensionless t/d is used to replace t and $\alpha(t/d)$ is used to describe the relationship between σ and t/d in a manner similar to that of the strain-hardening relation $\sigma = C\varepsilon^n$. Accordingly, $\alpha(t/d) = 0$, defined at $t/d = 0$ (as $d \to \infty$), the equation represents the case of a single crystal, and then $\sigma(0) = \sigma_0$. When $\alpha(t/d) = 1$, at $t/d \to (t/d)_c$, it denotes the critical condition that distinguishes the macroscale from microscale in the flow stress of the simple tension. Therefore, $\sigma[(t/d)_c] = \sigma_\infty$.

For the earlier assumption, the following is assumed:

$$\sigma_0 = C_0\varepsilon^a \text{(for the case of single crystal)} \tag{2.38}$$

And

$$\sigma_\infty = C_\infty \varepsilon^b \text{ (for the case of polycrystal)} \tag{2.39}$$

where C_0 and C_∞ are material parameters. Accordingly, the earlier two equations are similar to the equations for the power-law hardening of polycrystalline materials. By further assumption that $a = b = n$ by Leu et al. [25], the flow stress becomes

$$\sigma(t/d, \varepsilon) = C_0 \varepsilon^n + \alpha(t/d)\left(C_\infty \varepsilon^n - C_0 \varepsilon^n\right)$$

$$= \left[C_0 + \alpha(t/d)(C_\infty - C_0)\right]\varepsilon^n \tag{2.40}$$

In the earlier equation, it is noticed that the parameter $\alpha(t/d)$ plays an important role in the modeling of tensile flow stress of metals in microscale. Based on the relationship between the flow stress and the ratio t/d, as shown in Figure 2.17, the parameter $\alpha(t/d)$ can be assumed as a hyperbolic function in the following:

$$\alpha(t/d) = \tanh\left[C_a(t/d) / (t/d)_c\right] \tag{2.41}$$

where C_a is a control factor, which is generally set to 3.0 in practice, because $\alpha(t/d) = \tanh(C_a) = \tanh(3.0) = 0.995 \approx 1.0$ at $t/d = (t/d)_c$.

In the elementary analysis, the pileup theory of dislocations is employed to find the critical value $(t/d)_c$. When the shear stress of the mth grain decreases to the friction stress of a single crystal (the limiting stress) τ_i, these interactive grains are defined as an affected zone, as shown in Figure 2.18. Thus, these m grains comprise an affected zone and exhibit the following relation:

$$\tau_i = \beta^m \left(\tau_{rs}\sqrt{2}\right)\left(\frac{1}{2}\right)^{m/2} \tag{2.42}$$

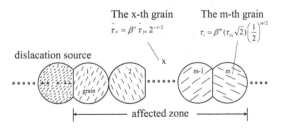

FIGURE 2.18
The model of the affected zone [25].

In this situation, the resolved shear stress τ_{rs} and the friction stress (limiting stress) τ_i can be represented by the Hall–Petch relation, a strength relation for a polycrystal, with grain size d, in the following:

$$\tau_{rs} = \tau_i + \tau_s = \tau_i + \frac{k}{\sqrt{d}} \qquad (2.43)$$

From Eqs. (2.42) and (2.43), the number of grains forming an affected zone can be determined as

$$m = \ln\left(\frac{\tau_i}{\sqrt{2}\left(\tau_i + k\sqrt{d}\right)}\right) \Big/ \ln\left(\frac{\beta}{\sqrt{2}}\right) \qquad (2.44)$$

According to the symmetry of the microstructure of grains, the total number in the affected zone, given by the definition of the critical value $(t/d)_c$ herein, can be defined as

$$\left(t/d\right)_c = 2m + 1 \qquad (2.45)$$

which incorporates the affected grains on both sides of the source grain of dislocation. The affected zone seems to be a sphere that is filled with the affected grains, and the central one is the source grain of the dislocation. The critical value $(t/d)_c$ can be represented in terms of process parameters.

$$\left(t/d\right)_c = 2\ln\left(\frac{\tau_i}{\sqrt{2}\left(\tau_i + k\sqrt{d}\right)}\right) \Big/ \ln\left(\frac{\beta}{\sqrt{2}}\right) + 1 \qquad (2.46)$$

2.4.4 Modified Hall–Petch Model

In Hall–Petch model (Eq. 2.9), the second term is related to the contributions from the grain boundary to the total flow stress. For the case of a single crystal, however, there are no neighboring grains to resist the propagation of dislocations, and consequently, there would be no effect induced by the grain boundary. The internal grain boundary length per area (GBi/area) is decreased with the size of specimen, and eventually becomes 0 for a single crystal. This is explained by Gap et al. [26] in Figure 2.19 for the ideal case of square-shaped grains and specimen cross section. As the number of grains (n) increases, GBi/area is increased from 0 to 2 for the square grains.

Hence, for a given grain size, when the specimen size decreases from a polycrystal to single crystal, the effects induced by the grain boundary will diminish. The assumption may be represented in the following form:

$$\sigma(\varepsilon) = M\tau_R(\varepsilon) + \frac{k(\varepsilon)}{\sqrt{d}}\beta \qquad (2.47)$$

where $0 \leq \beta \leq 1$; and β is 1 for polycrystal and about 0 when the specimen size is reduced to a single crystal.

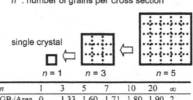

n^2: number of grains per cross section

n	1	3	5	7	10	20	∞
GBi/Area	0	1.33	1.60	1.71	1.80	1.90	2

FIGURE 2.19

Illustration of the reduction of internal grain boundary length per area (GBi/area) with miniaturization [26].

A comparison between the magnitude of the flow stress $\sigma(\varepsilon)$ and the Hall–Petch constant $\sigma_0(\varepsilon)$ supports that M is dependent on the specimen size scale. Since the second term in Eq. (2.48) is positive for the material, the value of M decreases as the number of grain (n) value decreases. In other words, the value of $\sigma_0(\varepsilon)$ is not a constant but varies with the ratio of the feature and grain sizes. Therefore, the following equation is proposed to describe the feature size effect on M.

$$\sigma_0(\varepsilon) = M^\alpha \tau_R(\varepsilon) \qquad (2.48)$$

where ($M^\alpha \geq 2, \alpha \leq 1$); and $\alpha = 1$ for polycrystal.

Combining Eqs. (2.48) and (2.49), a scaling equation including both the feature and grain size effects is derived as

$$\sigma(\varepsilon) = M^\alpha \tau_R(\varepsilon) + \frac{k(\varepsilon)}{\sqrt{d}}\beta \qquad (2.49)$$

where ($M^\alpha \geq 2, \alpha \leq 1, 0 \leq \beta \leq 1$); $\beta = 0$ for single crystal material; and $\alpha = \beta = 1$ for polycrystal one. It is also important to note that the limitation and assumption of the model. The model assumes that the grains are generally equiaxed, and thus the material behavior is assumed to be homogeneous. The model may be applied to deformed parts that are fully polycrystal and to the parts with only several grains. As the part size approaches the size of single grain, the texture of the grain, rather than the averaged effect from the slip systems, dominates the behavior.

2.5 Grain Statistics and Orientation Effect

2.5.1 Surface Roughening and Scattering

The ideal behavior of polycrystalline metal is isotropic since the ratio of specimen size to grain size is large and different grains with different sizes,

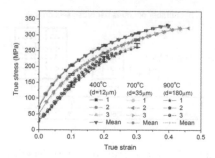

FIGURE 2.20
The flow stress curves of pure copper with different grain sizes [32].

shapes, and orientations are evenly and randomly distributed within the specimen. When the grain size is increased and the specimen size is kept constant, there will be only a few grains in the specimen, and the number of microstructural features is decreased. The even distribution of different grains no longer exists. Each grain plays a significant role to the overall material deformation behavior. The combined properties of grains determine the properties of specimen. In addition, different crystallographic orientations of neighboring grains lead to inhomogeneous deformation and the grains with different sizes, shapes, and orientations in different specimens result in the scatter of the measured material properties.

In Figure 2.20, it is found that the scattering effect of the obtained flow stress curves increases with the decreasing number of grains in the testing materials. It can also be noticed that the scatter of experimental data increases with the annealing temperature, indicating that the repeatability of experimental data becomes worse with the increase of grain size. This is because there are fewer grains in the deformation region as the grain size increases. The influence of individual grains on the flow stress thus becomes more prominent with the decrease of grain boundary density. As a consequence, the uneven distribution of individual grains with different sizes, shapes, and orientations in the deformation region leads to the significant inhomogeneous deformation and scattered material properties.

The interaction of the deformations among surface grains with different orientations makes the surface of the deformed micropart rougher. Figure 2.21 shows the scanning electron microscope images of the surface topographies of the compressed pure copper specimens under the same magnification ratio. The surface roughness corresponding to the specimen size increases with the decrease of specimen size. The plastic deformation is accommodated by slip bands, which increase the surface roughness microscopically. Furthermore, the surface grains are not strongly constrained, and the grain boundary sliding can perform easily in the deformation process. The strain incompatibility among grains further causes the inhomogeneous

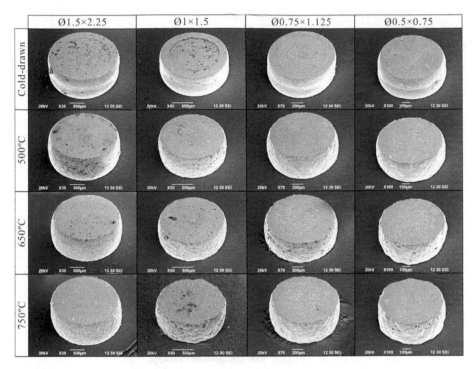

FIGURE 2.21
Surface topographies of the compressed pure copper specimens [51].

deformation and the free surface grains move normal to the surface. This leads to the occurrence of surface roughening.

On the other hand, quantitative evaluation of size effect can be done via the measurement of the surface roughness of the deformed specimens with different thicknesses and different grains by using a profilometer. The measured surface roughness of pure copper specimens after tensile tests are shown in Figure 2.22. The surfaces of specimens are roughened after deformation and are shown to increase significantly with grain size. The reason for the increase of surface roughness could be attributed to the uneven deformation of surface grains. With the increase of grain size, the individual grains, especially the surface grains, become less restricted due to the decrease of grain boundary density and more free surfaces of surface grains. Considering that the orientations and structures of individual grains are randomly distributed, the inhomogeneous and uneven deformations of surface grains become more significant, leading to the increase of surface roughness with grain size upon deformation.

The similar results of the increase of surface roughness with deformation and grain size can also be found in other studies by using steel, copper, and aluminum alloys [27–30]. From the experimental results, it is proved that the surface roughening effect has a close relationship with the specimen

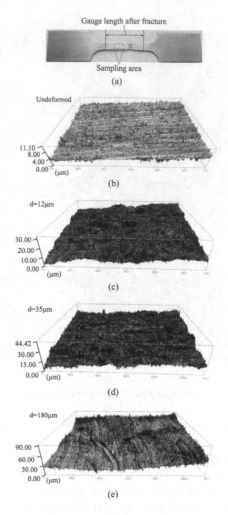

FIGURE 2.22
Surface microtopographies of specimens with different grain sizes before and after tension: (a) the schematic of testing area; (b) before deformation; (c) $d = 12\,\mu m$; (d) $d = 35\,\mu m$; (e) $d = 180\,\mu m$ [32].

size, grain size, and crystallographic texture. Such an effect influences not only the quality of microformed parts but also the production cost. Since the increase of surface roughness would initiate strain localization, it could result in cracks in the microformed parts. In addition, it also accelerates die wear due to the increase of interfacial friction. The size effect on surface roughness thus needs to be considered in microforming system design.

2.5.2 Constitutive Model with Orientation Effect

Since polycrystalline materials can be treated as an aggregate of many randomly distributed grains with different orientations, the properties of

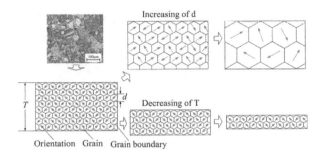

Increasing of d

Decreasing of T

Orientation Grain Grain boundary

FIGURE 2.23
Schematic description of the grain orientation effect [32].

materials are then determined by a collective function of each individual grains. In macroscale, there are a number of grains located in the section of materials. Since the orientations and shapes of grains are stochastically distributed, the material thus demonstrates uniform properties. On the other hand, as illustrated in Figure 2.23, if the grain size becomes closer to the feature size of workpiece and the deformation scale is scaled down to meso- and microlevel, there are only a few grains in the deformation section. As a result, the orientation and shape of each individual grain play a more significant role in the overall performance of materials. In addition, the grains, especially the surface ones, are also less constrained. The collective material behavior thus performs higher uncertainty due to the random properties of individual grains. This can be an intrinsic reason for the size effect caused by surface grains.

In order to describe this effect, the traditional Hall–Petch equation is modified by considering the contribution of each individual grains with different shapes, sizes, and orientations. The contribution of each grain is weighted according to the volume fraction f_i in the polycrystalline material. The total flow stress of specimen can be denoted as

$$\sigma = \sum_{i=1}^{n} f_i M_i \left(\tau_i' + K_i' d^{-\frac{1}{2}} \right) \tag{2.50}$$

In Eq. (2.50), the overall flow stress is considered to be composed of n types of grains with different orientations.

The fraction f_i in the polycrystalline material changes with deformation [31]. The scattered inverse pole figure grains of pure copper before and after deformation along the tensile axis are presented in Figure 2.24. It should be noticed that the grain orientations evolve in the uniaxial test according to the electron backscatter diffraction (EBSD) observations. The fraction of <110> orientated grains decreases prominently while the fraction of grains with the orientation of <111> increases significantly.

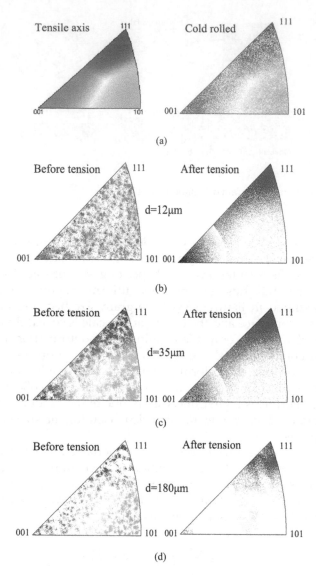

FIGURE 2.24

Scattered inverse pole figures of copper specimens with different grain sizes: (a) as-received; (b) $d = 12\,\mu m$; (c) $d = 35\,\mu m$; (d) $d = 180\,\mu m$ [32].

To represent the evolution of grain orientations, the changing fraction of grains with a specific orientation can be assumed to follow a linear function of strain throughout the uniaxial deformation for simplification:

$$f_i(\varepsilon) = f_{ib} + \frac{\left(f_{if} - f_{ib}\right)}{\left(\varepsilon_{if} - \varepsilon_{ib}\right)}\varepsilon \qquad (2.51)$$

TABLE 2.1

Volume Fraction of Grains with Different Orientations [32]

Grain size (μm)	State	<110>	<100>	<111>
12	Before tension	0.44	0.23	0.33
	After tension	0.06	0.25	0.69
35	Before tension	0.48	0.15	0.37
	After tension	0.08	0.19	0.73
180	Before tension	0.63	0.10	0.27
	After tension	0.07	0.007	0.923

In Eq. (2.51), f_{ib} is the fraction of the grains with the ith orientations and a smaller deformation of ε_{ib}, while f_{if} is the fraction of the grains with a larger deformation of ε_{if}. With the combination of Eqs. (2.50) and (2.51), the constitutive model can represent the scattering effect and orientation evolution of metallic materials in microforming process.

According to the EBSD results of pure copper specimens before and after tensile deformation in Figure 2.24, the fractions of the orientations of three main groups including <110>, <100>, and <111> are listed in Table 2.1.

When only the three orientations of pure copper materials, i.e., <110>, <100> and <111>, are considered, the total flow stress of material can thus be determined as

$$\sigma = f_1 M_1 \left(\tau_1' + K_1' d^{-\frac{1}{2}} \right) + f_2 M_2 \left(\tau_2' + K_2' d^{-\frac{1}{2}} \right) + f_3 M_3 \left(\tau_3' + K_3' d^{-\frac{1}{2}} \right) \quad (2.52)$$

In Eq. (2.52), f_i and M_i are the volume fraction and orientation factor of grain group i with a specific orientation, respectively. The footnote $i = 1, 2, 3$, represents the three main orientations as discussed earlier. Assuming τ_i' and K_i' follows the power and exponential function of strain, the following equation can be obtained:

$$\sigma = f_1 M_1 \left(A_1 \varepsilon^{m_1} + B_1 \varepsilon^{n_1} d^{-\frac{1}{2}} \right) + f_2 M_2 \left(A_2 \varepsilon^{m_2} + B_2 \varepsilon^{n_2} d^{-\frac{1}{2}} \right)$$

$$+ f_3 M_3 \left(A_3 \varepsilon^{m_3} + B_3 \varepsilon^{n_3} d^{-\frac{1}{2}} \right) \quad (2.53)$$

where A_i, B_i, m_i, and n_i are material constants.

The fractions of the grain groups with three orientations are proportionally extrapolated based on the experimental results. With regard to the experimental investigations shown in Table 2.1, the changing fraction of grains with a specific orientation can be assumed to follow a linear function of strain throughout the uniaxial deformation for simplification in the following:

$$
\begin{cases}
f_1(\varepsilon) = f_{1i} + \dfrac{(f_{1f} - f_{1i})}{(\varepsilon_{1f} - \varepsilon_{1i})}\varepsilon \\[2ex]
f_2(\varepsilon) = f_{2i} + \dfrac{(f_{2f} - f_{2i})}{(\varepsilon_{2f} - \varepsilon_{2i})}\varepsilon \\[2ex]
f_3(\varepsilon) = f_{3i} + \dfrac{(f_{3f} - f_{3i})}{(\varepsilon_{3f} - \varepsilon_{3i})}\varepsilon
\end{cases}
\tag{2.54}
$$

In Eq. (2.54), f_{1i}, f_{2i}, and f_{3i} are fractions of the grains with the orientations of <110>, <100>, and <111> before deformation, respectively, while f_{1f}, f_{2f}, and f_{3f} are the fractions after deformation. Figure 2.25 shows the comparison of the analytical and experimental flow stress of pure copper specimen with the grain size of 12, 35, and 180 μm. It can be observed that the calculated results with the consideration of grain orientation development agree with the experimental results.

2.5.3 Simulation with Consideration of Grain Statistics

To simulate the microforming process considering grain statistics size effect, three issues need to be considered: (a) establishment of virtual grains in metallic materials, (b) mechanical behaviors of grains with various orientations, and (c) repeated case study for the scattering effect. For the first issue, Voronoi theory [33–35] is widely used to establish the geometrical model of virtual grains, and the orientations of grains are defined statistically according to the obtained fractions of different orientations from the EBSD observations. For the second issue, the mechanical properties are defined for each grain according to its orientation based on the predictive results of the established model. After that, three different structures are generated for each grain size condition to verify the repeatability of simulation. The simulation for each structure was repeated many times (about 20 times) using different grain orientation mapping solutions. After that, each grain section was meshed as shown in Figure 2.26. CPS4R elements were employed in the meshing process. The grain sections were divided randomly into three groups corresponding to the three grain orientations using the Matlab program. The grain sections were then given with the calculated plastic properties corresponding to the three grain orientations during deformation. Twenty different mapping (or grouping) solutions were used for each grain structure. One of the mapping solutions is shown in Figure 14. A displacement load was then applied by stretching one side of the sample in the longitude direction while fixing the other. In addition, the symmetric boundary condition was implemented at the lower surface.

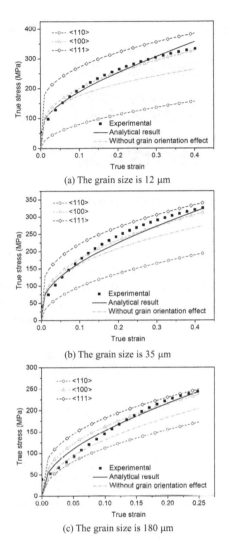

FIGURE 2.25
Comparison of the analytical and experimental results of pure copper specimen with various grain sizes [32].

Based on the virtual grain model, simulations of tensile test were carried out, and the surface topography of specimen for various grain sizes is shown in Figure 2.27. To calculate the roughness, the heights of all nodes in the upper surface were recorded as $h_i(i = 1, 2,...,n)$. The average height was calculated as $\bar{h}$. Hence the surface roughness could be estimated according to $R_a = \sum_{i=1}^{n} |h_i - \bar{h}| / n$. Both the average value and the scatter of surface roughness are shown to increase with grain size according to the simulation results.

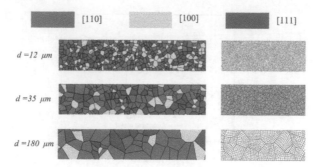

FIGURE 2.26
The FE models of the materials with different grain sizes [32].

This is because the mechanical properties of individual grains are different from each other, leading to significant inhomogeneous deformation. In addition, the uneven deformation can be clearly observed. With the increase of grain size, the number of grains in the deformation area decreases. Therefore, the individual grains are less restricted by each other and the inhomogeneous deformation thus becomes more severe, resulting in the increase of

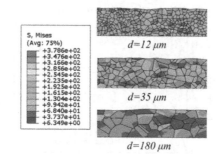

(a) The simulated free surface morphology after deformation

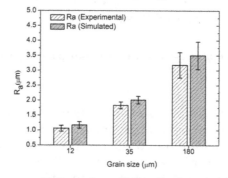

(b) The roughness of deformed surface

FIGURE 2.27
Surface topography analysis of the FE results [32].

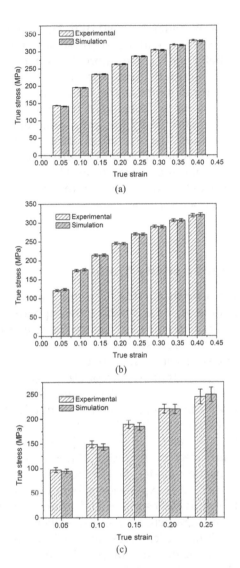

FIGURE 2.28
Comparison of the simulative and experimental stress–strain relations for the specimen with
various grain sizes: (a) $d = 12\,\mu m$, (b) $d = 35\,\mu m$, and (c) $d = 18\,\mu m$ [32].

surface roughness. The finite element (FE) simulation thus provides an intuitive and persuasive demonstration of the surface roughness size effect.

The true stress–strain curves for different grain sizes were also obtained
as shown in Figure 2.28. The simulated average flow stresses agree with the
experimental results under different strain and grain size conditions. The
scatter of simulated results also increases with the grain size, which is in
accordance with the experimental observation. This is because the properties

of individual grains were defined statistically. The number of grains in the deformation area is decreased with the increase of grain size. Therefore, the overall deformation behavior becomes more dependent on the property and deformation of individual grain. Since the properties of individual grains demonstrate significant randomness, the scatter of data increases with the grain size. Therefore, it can be concluded that the developed model can describe the flow stress affected by the grain orientation effect reasonably under different grain sizes. The size effect affected deformation in meso- and microforming can be represented and approximated by an intuitive and practical manner via combining this model and FE tools. This work thus provides a basic and effective tool for understanding and analyzing the intrinsic reason of size effect in microforming.

2.6 Strain Gradient Plasticity Model

In the conventional plastic theory, the flow stress in the forming process is described by the local flow stress, which only depends on the local strain, strain rate, etc. Different from the local theory, the effects of strain gradients on the material behaviors are also taken into account and nonlocal theory are developed to represent feature size effects.

Generally, there are two frameworks of strain gradient plasticity theories based on the order of the equivalent stress and strain involved. The first framework is Mindlin's higher-order continuum theories. Fleck and Hutchinson [36] used the dislocation theory to develop their formulation of strain gradient plasticity. Later, Gao et al. [37] and Huang et al. [38,39] proposed the mechanism-based strain gradient (MSG) plasticity theory derived from the Taylor dislocation model. These theories have shown reasonable agreement with the microscale experiments. However, the actual theory was built by replacing the equivalent stresses and strains with the ones of higher order that contain strain gradient terms, respectively. The other framework does not contain the higher-order stress, and the effect of the strain gradient is taken into account by the incremental plastic moduli. It includes the work of Acharya [40] and Evers et al. [41]. Later on, Huang et al. [42,43] proposed a conventional theory of MSG (CMSG) plasticity. The model preserves the structure of conventional plasticity theories. The governing equations and boundary conditions of the CMSG model are the same as the conventional continuum theories, while the constitutive model contains the term of plastic strain gradient. Taylor dislocation model and CMSG plasticity model are introduced in detail in the following sections.

2.6.1 Taylor Dislocation Model

In the Taylor dislocation model, the shear flow stress in terms of the dislocation density is expressed as [43]

$$\tau = \alpha\mu b\sqrt{\rho_S + \rho_G} \tag{2.55}$$

where τ is the shear flow stress, μ is the shear modulus, b is the magnitude of Burgers vector, α is an empirical coefficient around 0.3, and ρ_s and ρ_G are the densities of statistically stored and geometrically necessary dislocations, respectively. The tensile flow stress σ_{flow} can be expressed by $\sigma_{flow} = M\tau$, where M is the Taylor coefficient. $M = 3.06$ stands for an FCC crystal and body-centered cubic crystal with <110> slip plane [44].

The density of geometrically necessary dislocation ρ_G is related to the gradient of plastic strain by $\rho_G = \bar{r}\eta_p/b$, where $\bar{r}$ is the Nye factor and $\bar{r} = 1.9$ for FCC crystals, η_p is the equivalent plastic strain gradient. The density of statistically stored dislocations ρ_s is determined by the equation $\sigma_{ref} f_p(\varepsilon_p)/M = \alpha\mu b\sqrt{\rho_S}$ in uniaxial tension, where σ_{ref} is a reference stress, $f(\varepsilon^p)$ is a function of equivalent plastic strain representing the stress and plastic strain relation, and ε^p is the plastic strain in uniaxial tension.

According to the Taylor dislocation model in Eq. (2.55), the flow stress is given in terms of the strain and plastic strain gradient as follows:

$$\sigma_{flow} = \sigma_{ref}\sqrt{\left[f\left(\varepsilon^p\right)\right]^2 + l\eta^p} \tag{2.56}$$

The earlier equation is the Taylor dislocation model considering strain gradient. l is the intrinsic material length related to strain gradient plasticity and is in the order of micrometers [45].

$$l = \frac{M^2\bar{r}\alpha^2\mu^2 b}{\sigma_{ref}^2} = M^2\bar{r}\alpha^2\left(\frac{\mu}{\sigma_{ref}}\right)^2 b \tag{2.57}$$

$$\eta^p = \int \dot{\eta}^p \, dt, \quad \dot{\eta}^p = \sqrt{\frac{1}{4}\dot{\eta}_{ijk}^p \dot{\eta}_{ijk}^p} \tag{2.58}$$

$$\dot{\eta}_{ijk}^p = \dot{\varepsilon}_{jk,i}^p + \dot{\varepsilon}_{ik,j}^p - \dot{\varepsilon}_{ij,k}^p, \quad \dot{\varepsilon}_{jk,i}^p = \frac{\partial\varepsilon_{jk}^p}{\partial i} \tag{2.59}$$

where $\varepsilon_{jk}^p = \int \dot{\varepsilon}_{jk}^p dt$ is the plastic strain tensor.

The power-law viscoplastic model proposed by Kok et al. [46] can be expressed as

$$\dot{\varepsilon}^p = \dot{\varepsilon}\left[\frac{\sigma_e}{\sigma_{flow}}\right]^m \tag{2.60}$$

where $\dot{\varepsilon} = \sqrt{\frac{2}{3}\dot{\varepsilon}_{ij}'\dot{\varepsilon}_{ij}'}$ is the equivalent strain rate, $\sigma_e = \sqrt{\frac{3}{2}\sigma_{ij}'\sigma_{ij}'}$ is the equivalent stress, $\dot{\varepsilon}_{ij}'$ and σ_{ij}' are the deviatoric components of strain rate and stress, respectively, and m is the rate-sensitivity exponent. Huang et al. [42] demonstrated that Eq. (2.60) is valid for power-law hardening materials ($m \geq 20$).

Replacing the flow stress in Eq. (2.60) by the Taylor dislocation model, the plastic strain rate can be expressed as

$$\dot{\varepsilon}^p = \dot{\varepsilon} \left[\frac{\sigma_e}{\sigma_{ref} \sqrt{\left[f\left(\varepsilon^p\right) \right]^2 + l\eta^p}} \right]^m \tag{2.61}$$

2.6.2 Constitutive Model of CMSG

Based on the MSG plastic theory [37,38], Huang and his colleagues [42] proposed a CMSG plasticity, preserving the requirements of classical continuum plasticity without additional conditions involving high-order stress and strain.

The volumetric strain rate $\dot{\varepsilon}_{kk}$ and deviatoric strain rate $\dot{\varepsilon}'_{ij}$ in CSMG plasticity are related to the stress rate in the same way as in classical plasticity [42], i.e.

$$\dot{\varepsilon}_{kk} = \frac{\dot{\sigma}_{kk}}{3K} \tag{2.62}$$

$$\dot{\varepsilon}'_{ij} = \frac{1}{2\mu} \dot{\sigma}'_{ij} + \frac{3\dot{\varepsilon}^p}{2\sigma_e} \sigma'_{ij} \tag{2.63}$$

By rearranging Eqs. (2.61)–(2.63), the strain rate and the stress rate in terms of strain rate and strain gradient can be obtained by the following expression:

$$\begin{cases} \dot{\varepsilon}_{ij} = \frac{\dot{\sigma}'_{ij}}{2\mu} + \frac{\dot{\sigma}_{kk}}{9K} \delta_{ij} + \frac{3\dot{\varepsilon}}{2\sigma_e} \left[\frac{\sigma_e}{\sigma_{ref} \sqrt{f^2\left(\varepsilon^p\right) + l\eta^p}} \right]^m \sigma'_{ij} \\[3em] \dot{\sigma}_{ij} = K\dot{\varepsilon}_{kk}\delta_{ij} + 2\mu \left\{ \dot{\varepsilon}'_{ij} - \frac{3\dot{\varepsilon}}{2\sigma_e} \left[\frac{\sigma_e}{\sigma_{ref} \sqrt{f^2\left(\varepsilon^p\right) + l\eta^p}} \right]^m \sigma'_{ij} \right\} \end{cases} \tag{2.64}$$

where $\dot{\sigma}'_{ij}$ is the deviatoric stress rate and K is the bulk modulus. The constitutive relation expressed earlier represents the strain gradient effect without containing the higher-order stress. Accordingly, the equilibrium equations, boundary conditions, and algorithms used in classical mechanics can be conveniently implemented.

2.6.3 Simulation of Microdeformation Based on CMSG Model

Currently, most commercial FE method software is based on the classical plastic theory, which is quite mature in solving the problems arising from the conventional forming process. However, they cannot provide accurate simulation of meso- and microforming because the size effects are not taken into consideration in numerical simulation. Generally, to simulate microforming based on the CMSG model, the interface of user-defined material (UMAT) subroutine provided by commercial ABAQUS/Standard software is applied. Figure 2.29 shows the flow chart of the UMAT based on the CMSG plasticity model.

The effort beyond that in classical plasticity is the evaluation of the plastic strain gradient. It should be accomplished by interpolating the plastic strain increment for the Gaussian integration points in each element. It is generally

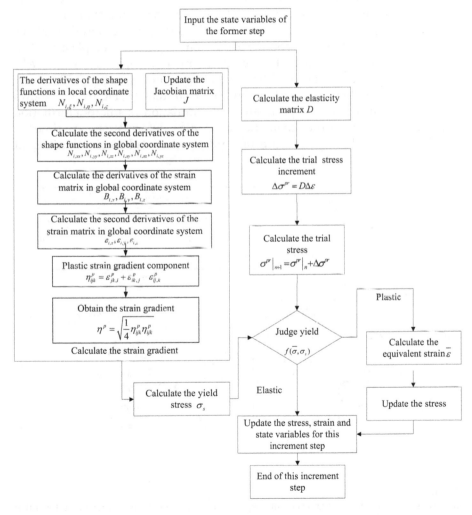

FIGURE 2.29
Flow chart of the UMAT [49].

supposed that the elastic strain is much smaller than the plastic strain during the forming process. Thus, the plastic strain approximately equals to the total strain. The assumption was adopted by Swaddiwudhipong et al. [47], Fleck and Hutchinson [48] as well. Therefore, the plastic strain gradient equals to strain gradient, and the calculation procedure of the equivalent plastic strain gradient could be simplified.

The gradient of plastic strain increment is determined via the differentiation of the shape function in the isoparametric space. First of all, the derivatives of shape functions in local coordinate system $N_{i,\xi}$, $N_{i,\eta}$, $N_{i,\zeta}$ could be established, and the Jacobian matrix J was updated according to the state variables of the former increment step. The second derivatives of shape functions in the global coordinate system $N_{i,xx}$, $N_{i,yy}$, $N_{i,zz}$, $N_{i,xy}$, $N_{i,xz}$, $N_{i,yz}$ could be calculated. Therefore, the derivatives of the strain matrix in global coordinate system $B_{i,x}$, $B_{i,y}$, $B_{i,z}$ could be obtained. By calculating the second derivatives of the strain matrix in global coordinate system, $\varepsilon_{i,x}$, $\varepsilon_{i,y}$, $\varepsilon_{i,z}$, the plastic strain gradient component could be further determined via $\eta^p_{ijk} = \varepsilon^p_{jk,i} + \varepsilon^p_{ik,j} - \varepsilon^p_{ij,k}$. Finally, the strain gradient could be calculated according to Eq. (2.59).

Based on the earlier described method of FE simulation, Figure 2.30 shows the comparison between experiment and simulation of microdeformation [49]. Figure 2.30a gives the schematic of the microdeformation experiments, and Figure 2.30b displays two sets of rigid punches and dies with different

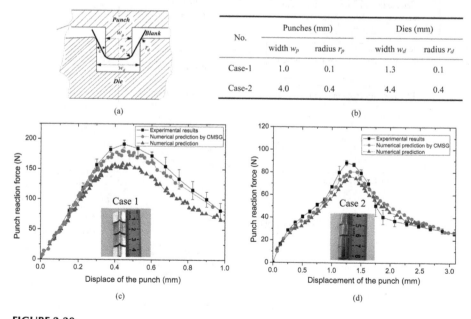

No.	Punches (mm)		Dies (mm)	
	width w_p	radius r_p	width w_d	radius r_d
Case-1	1.0	0.1	1.3	0.1
Case-2	4.0	0.4	4.4	0.4

(a)   (b)

(c)   (d)

FIGURE 2.30
Comparison of the force–displacement response for numerical and experimental results: (a) schematic of the microdeformation experiments, (b) two sets of rigid punches and dies, (c) and (d) pure reaction force versus displacement of the punch [49].

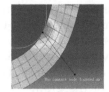

No.	$\left[f(\varepsilon^p)\right]^2$	$l\eta^p$	$\dfrac{l\eta^p}{\left[f(\varepsilon^p)\right]^2 + l\eta^p}$
Case-1	0.154	0.033	17.6%
Case-2	0.073	0.006	8.1%

FIGURE 2.31
The contact point and the strain gradient contribution to the flow stress [49].

dimensions of microchannels. It can be seen that the width of microchannel for cases 1 and 2 are 1.0 and 4.0 mm, respectively. The obtained displacement–force relations from experiments and simulations are separately displayed in Figure 2.30c and d for the two cases. It is clearly found that the simulation results using the CMSG plastic theory agree better with the experimental results compared with those of the classic plastic theory, especially for the minimized case 1 (1.0 mm wide channel), in which the equivalent plastic strain gradient is much more important.

According to the constitutive model designated by Eq. (2.56), it is found that the flow stress is affected not only by strain but also by strain gradient. The values of $[f(\varepsilon^p)]^2$ and $l\eta^p$ determine whether the strain gradient affects the flow stress or not. If the values of $[f(\varepsilon^p)]^2$ and $l\eta^p$ are in the same magnitude, such as in meso- and microforming, the involvement of strain gradient thus affects the flow stress. On the contrary, if the value of $l\eta^p$ is not comparative with $[f(\varepsilon^p)]^2$, such as in macroscale forming, the influence of strain gradient can be ignored. Figure 2.31 shows the contact point and the strain gradient contribution to the flow stress [49]. The contribution of strain gradient to the flow stress increases with the decrease of the tool dimension from 4.0 to 1.0 mm. Actually, the further increase of strain grain gradient can be caused by the decrease of tool geometrical parameters and blank thickness.

2.7 Summary

In meso- and microforming, the similarity theory is commonly used for an easy evaluation of the size effect on the mechanical response of metals. When the geometrical parameters of a traditional macroforming system and a microforming system are similar, the similarity difference and similar precision are generally used as similarity benchmark indicators. When the similarity difference or similar precision does not equal to 0, it means that a size effect exists. With the increasing product miniaturization, two different types of size effects on mechanical response of materials generally occur. One is the grain size effect and the other is geometric size effect. Generally speaking, in meso- and microforming, the flow stress of metallic materials decreases with the increase of grain size, and the decrease of feature size or the ratio of the specimen size to the grain size (t/d).

The grain size effect can be explained by the grain boundary strengthening, which is actually caused by the pileup of dislocations. Hall–Patch model is commonly used for the representation of the grain size effect in meso- and microforming. On the other hand, for geometric size effect, the physical mechanisms and relatively theoretical modeling are still the key issues to be addressed in microforming of metallic materials. Among these studies, four types of theories get lots of attention, and they are surface layer model, grain boundary model, crystal plastic model as well as modified Hall–Petch model. Although these models are established from different perspectives, it is believed that with the decreasing size scale, the material deformation behavior changes from polycrystal mode in macroscale to single crystal mode in microscale in the extreme scenario. For the meso- and microscales, the single crystal model is the lower bound and the polycrystal is the upper bound for modeling of material deformation.

When the number of microstructural features is decreased and there are only a few grains in the specimen, the even distribution of different grains no longer exists. Different crystallographic orientations of grains lead to inhomogeneous deformations and scatter of the measured material properties. For the scattering problems, the flow stress is modeled as a combination of the flow stress curves of single grains with various orientations. To consider grain statistics size effect in the simulation of microforming process, Voronoi theory is widely used to establish the geometrical model of virtual grains, and the orientations of grains are defined statistically according to the obtained fractions of different orientations from the EBSD observations.

Furthermore, the effects of strain gradients on the material behaviors are also taken into account and the nonlocal theory is developed to represent size effects related with the process conditions. The most widely used models are Taylor dislocation model and CMSG model. The CMSG model is relatively easy for implementation in finite element simulation of microforming due to not using the high-order stress. Based on the CMSG model, the numerical simulation of microforming with the consideration of the effect of strain gradients is also introduced in this chapter.

References

1. U. Engel, R. Eckstein, Microforming—From basic research to its realization, *Journal of Materials Processing Technology* 125–126 (2002) 35–44.
2. K.F. Zhang, L. Kun, Classification of size effects and similarity evaluating method in micro forming, *Journal of Materials Processing Technology* 209(11) (2009) 4949–4953.

3. E. Hall, The deformation and ageing of mild steel: III discussion of results, *Proceedings of the Physical Society. Section B* 64(9) (1951) 747.
4. R.W. Armstrong, I. Codd, R.M. Douthwaite, N.J. Petch, The plastic deformation of polycrystalline aggregates, *Philosophical Magazine* 7(73) (1962) 45–58.
5. M. Geiger, A. Messner, U. Engel, Production of microparts—Size effects in bulk metal forming, similarity theory, *Production Engineering* 4(1) (1997) 55–58.
6. T.A. Kals, R. Eckstein, Miniaturization in sheet metal working, *Journal of Materials Processing Technology* 103(1) (2000) 95–101.
7. J. An, Y.F. Wang, Q.Y. Wang, W.Q. Cao, C.X. Huang, The effects of reducing specimen thickness on mechanical behavior of cryo-rolled ultrafine-grained copper, *Materials Science and Engineering: A* 651 (2016) 1–7.
8. L.V. Raulea, L.E. Govaert, F.P.T. Baaijens, Grain and specimen size effects in processing metal sheets, *Advanced Technology of Plasticity* 2 (1999) 19–24.
9. L.V. Raulea, A.M. Goijaerts, L.E. Govaert, F.P.T. Baaijens, Size effects in the processing of thin metal sheets, *Journal of Materials Processing Technology* 115(1) (2001) 44–48.
10. W.L. Chan, M.W. Fu, Experimental studies and numerical modeling of the specimen and grain size effects on the flow stress of sheet metal in microforming, *Materials Science and Engineering: A* 528(25–26) (2011) 7674–7683.
11. S. Miyazaki, K. Shibata, H. Fujita, Effect of specimen thickness on mechanical properties of polycrystalline aggregates with various grain sizes. *Acta Metallurgica* 27(5) (1979) 855–862.
12. C.J. Wang, C.J. Wang, J. Xu, P. Zhang, D.B. Shan, B. Guo, Plastic deformation size effects in micro-compression of pure nickel with a few grains across diameter, *Materials Science and Engineering: A* 636 (2015) 352–360.
13. J.T. Gau, C. Principe, J. Wang, An experimental study on size effects on flow stress and formability of aluminm and brass for microforming, *Journal of Materials Processing Technology* 184(1–3) (2007) 42–46.
14. E. Schmid, W. Boas, *Plasticity of Crystals*, Chapman & Hall Ltd., London, 1950.
15. R.W. Armstrong, On size effects in polycrystal plasticity, *Journal of the Mechanics and Physics of Solids* 9(3) (1961) 196–199.
16. H. Mecking, U.F. Kocks, Kinetics of flow and strain-hardening, *Acta Metallurgica* 29(11) (1981) 1865–1875.
17. R.W. Armstrong, The influence of polycrystal grain size on several mechanical properties of materials, *Metallurgical and Materials Transactions B* 1(5) (1970) 1169–1176.
18. B. Clausen, T. Lorentzen, T. Leffers, Self-consistent modelling of the plastic deformation of fcc polycrystals and its implications for diffraction measurements of internal stresses, *Acta Materialia* 46(9) (1998) 3087–3098.
19. R.W. Armstrong, The yield and flow stress dependence on polycrystal grain size, in *Yield, Flow and Fracture of Polycrystals*, Editor T.N. Baker, Applied Science Pub., London, 1983, pp. 1–31.
20. L.F. Peng, X.M. Lai, H.J. Lee, J.H. Song, J. Ni, Analysis of micro/mesoscale sheet forming process with uniform size dependent material constitutive model, *Materials Science and Engineering A – Structural Materials Properties Microstructure and Processing* 526(1–2) (2009) 93–99.
21. M.A. Meyers, E. Ashworth, A model for the effect of grain-size on the yield stress of metals. *Philosophical Magazine* 46(5) (1982) 737–759.

22. H.H. Fu, D.J. Benson, M.A. Meyers, Analytical and computational description of effect of grain size on yield stress of metals, *Acta Materialia* 49(13) (2001) 2567–2582.
23. M. Geiger, S. Geibdorfer, U. Engel, Mesoscopic model – Advanced simulation of microforming processes, *Production Engineering*, 1 (2007) 79–84.
24. J.G. Liu, M.W. Fu, W.L. Chan, A constitutive model for modeling of the deformation behavior in microforming with a consideration of grain boundary strengthening, *Computational Materials Science* 55 (2012) 85–94.
25. D.K. Leu, Modeling of size effect on tensile flow stress of sheet metal in microforming, *Journal of Manufacturing Science and Engineering* 131(1) (2009) 011002.
26. G.Y. Kim, J. Ni, M. Koc, Modeling of the size effects on the behavior of metals in microscale deformation processes, *Journal of Manufacturing Science and Engineering* 129(3) (2007) 470–476.
27. D. Chandrasekaran, M. Nygårds, A study of the surface deformation behaviour at grain boundaries in an ultra-low-carbon steel, *Acta Materialia* 51(18) (2003) 5375–5384.
28. N. Tiesler, U. Engel, Microforming – Effects of miniaturization, in *8th International Conference on Metal Forming*, Rotterdam, A.A. Balkema, 2000 pp. 355–360.
29. C.J. Wang, C.J. Wang, B. Guo, D.B. Shan, Y.Y. Chang, Mechanism of size effects in microcylindrical compression of pure copper considering grain orientation distribution, *Rare Metals* 32(1) (2013) 18–24.
30. J. Xu, X.C. Zhu, D.B. Shan, B. Guo, T.G. Langdon, Effect of grain size and specimen dimensions on micro-forming of high purity aluminum, *Materials Science and Engineering: A* 646 (2015) 207–217.
31. J. Chen, W. Yan, B. Li, X.G. Ma, X.Z. Du, X.H. Fan, Microstructure and texture evolution of cold drawing <110> single crystal copper, *Science China-Technological Sciences* 54(6) (2011) 1551–1559.
32. L. Peng, A constitutive model for metal plastic deformation at micro/meso scale with consideration of grain orientation and its evolution, *Journal of Materials Processing Technology* (2017).
33. D. Stoyan, W.S. Kendall, J. Mecke, L. Ruschendorf, *Stochastic Geometry and Its Applications*, Vol. 2, Wiley, Chichester, 1995.
34. H.X. Zhu, S.M. Thorpe, A.H. Windle, The geometrical properties of irregular two-dimensional Voronoi tessellations, *Philosophical Magazine A* 81(12) (2001) 2765–2783.
35. A. Okabe, B. Boots, K. Sugihara, S.N. Chiu, *Spatial Tessellations: Concepts and Applications of Voronoi Diagrams*, Vol. 501, Wiley.com, New York, 2009.
36. N. Fleck, G. Muller, M. Ashby, J. Hutchinson, Strain gradient plasticity: Theory and experiment, *Acta Metallurgica et Materialia* 42(2) (1994) 475–487.
37. H. Gao, Y. Huang, W.D. Nix, J.W. Hutchinson, Mechanism-based strain gradient plasticity—I. Theory, *Journal of the Mechanics and Physics of Solids* 47(6) (1999) 1239–1263.
38. Y. Huang, Gao H., W.D. Nix, J.W. Hutchinson, Mechanism-based strain gradient plasticity—II. Analysis, *Journal of the Mechanics and Physics of Solids* 48(1) (2000) 99–128.
39. Y. Huang, Z. Xue, H. Gao, W.D. Nix, Z.C. Xia, A study of microindentation hardness tests by mechanism-based strain gradient plasticity, *Journal of Materials Research* 15(8) (2000) 1786–1796.

40. A. Acharya, J.L. Bassani, Lattice incompatibility and a gradient theory of crystal plasticity, *Journal of the Mechanics and Physics of Solids* 48(8) (2000) 1565–1595.
41. L.P. Evers, D.M. Parks, W.A.M. Brekelmans, M.G.D. Geers, Crystal plasticity model with enhanced hardening by geometrically necessary dislocation accumulation, *Journal of the Mechanics and Physics of Solids* 50(11) (2002) 2403–2424.
42. Y. Huang, S. Qu, K.C. Hwang, M. Li, H. Gao, A conventional theory of mechanism-based strain gradient plasticity, *International Journal of Plasticity* 20(4–5) (2004) 753–782.
43. H. Wang, K.C. Hwang, Y. Huang, P.D. Wu, B. Liu, G. Ravichandran, C.S. Han, H. Gao, A conventional theory of strain gradient crystal plasticity based on the Taylor dislocation model, *International Journal of Plasticity* 23(9) (2007) 1540–1554.
44. U. Kocks, The relation between polycrystal deformation and single-crystal deformation, *Metallurgical and Materials Transactions B* 1(5) (1970) 1121–1143.
45. B. Liu, Y. Huang, M. Li, K.C. Hwang, C. Liu, A study of the void size effect based on the Taylor dislocation model, *International Journal of Plasticity* 21(11) (2005) 2107–2122.
46. S. Kok, A.J. Beaudoin, D.A. Tortorelli, A polycrystal plasticity model based on the mechanical threshold, *International Journal of Plasticity* 18(5–6) (2002) 715–741.
47. S. Swaddiwudhipong, K.K. Tho, J. Hua, Z.S. Liu, Mechanism-based strain gradient plasticity in C0 axisymmetric element, *International Journal of Solids and Structures* 43(5) (2006) 1117–1130.
48. N.A. Fleck, J.W. Hutchinson, Strain gradient plasticity, *Advances in Applied Mechanics* 33 (1997) 295–361.
49. L.F. Peng, P.Y. Yi, P. Hu, X. Lai, J. Ni, Analysis of micro/mesoscale sheet forming process by strain gradient plasticity and its characterization of tool feature size effects, *Journal of Micro and Nano-Manufacturing* 3(1) (2015) 011006.
50. X.M. Lai, L.F. Peng, P. Hu, S.H. Lan, J. Ni, Material behavior modelling in micro/meso-scale forming process with considering size/scale effects, *Computational Materials Science* 43(4) (2008) 1003–1009.
51. W.L. Chan, M.W. Fu, B. Yang, Experimental studies of the size effect affected microscale plastic deformation in micro upsetting process, *Materials Science and Engineering A – Structural Materials Properties Microstructure and Processing* 534 (2012) 374–383.

3

Friction in Meso- and Microforming of Materials

3.1 Introduction

Friction has a great influence on manufacturing and product quality via its effect on loading force, change in the behavior of contact surface between materials and tools, and hindrance of the flow of materials in deformation-based manufacturing. It is thus one of the most important phenomena in metal-forming processes. Therefore, an in-depth understanding of friction phenomena and its determination of details become one of the most essential topics in metal-forming process design [1]. Unfortunately, there is no sufficient in-depth understanding of friction and its complex mechanisms, especially in the meso- and microforming of materials, in spite of the fact that friction has been known for many centuries. Friction behaviors are related to three main factors: the properties of materials, the profiles of tool–workpiece interfaces (roughness), and lubrication of interfaces. Some useful models including Coulomb, Constant, and General friction models have been developed based on the experimental observations and general description of friction in conventional forming processes [2]. However, the friction behaviors in meso- and microforming are different from those in macroscale due to the influence of size effect [3]. As a result, the conventional friction models may not be accurate when directly applied to meso- and microforming.

Therefore, many efforts have been carried out to study and determine the difference of friction in macro-, meso-, and microscale. Experiments such as double-cup extrusion, ring compression, and cylinder compression have been conducted in different scales, and it is found that friction increases greatly with the reduction of part size when the contacting surface is lubricated with liquid. However, in dry friction conditions, the phenomenon of size effect is not that significant. On the other hand, as individual grains play a more important role in meso- and microscale, the grain size effect is another essential issue in the investigation of interfacial friction. Although different friction tests were conducted to reveal the dependence of friction coefficient on grain size, a solid conclusion about the grain size effect has not yet been obtained. This is attributed to the complex surface deformation with friction. A remarkable work was done by Peng et al. [4], who studied the grain size effect by separating adhesion and plowing, two main friction mechanisms in deformation process, and found that grain size has an opposite effect on adhesive and plowing friction behaviors.

In modeling the geometry size effect on friction, the open–close lubricant pocket (O/CLP) model was widely used and further developed [5–7]. Engel [6] proposed a method to describe the geometry size effect by utilizing a scale factor representing the share of open pockets in the total lubricant pockets. Peng et al. [1] developed a uniform friction model to describe the friction behaviors from micro- to macroscale based on the assumption of O/CLPs, and implemented it into the finite element (FE) simulation. However, the lack of knowledge of the mechanisms behind friction obstacles the understanding of grain size effect on friction. Recently, Peng et al. [4] attributed the grain size effect on adhesive friction to the transformation from elastic to plastic deformation and the grain size effect on plowing friction to the transformation from intergranular to transgranular fracture. Based on the mechanism observed, friction models were developed for adhesive and plowing cases. In this chapter, both the geometry and grain size effects on friction behaviors and the mechanisms behind these are presented based on experiment, analysis, and simulation.

3.2 Size Effects on Interfacial Friction

Three kinds of size effects observed in experiments are described here, namely, asperity, geometry, and grain size effects. The asperity size effect refers to the evolution of friction mechanism with the change of surface asperity on the contacting surfaces. Surface asperity or roughness is the most crucial factor that affects friction behaviors. With the variation of surface asperity, the dominating mechanism of friction can be changed from adhesion to plowing, and the surface damage and friction force change significantly. Such kind of size effect is easy to observe and had been reported long time ago. Geometry size effect refers to the change of friction behavior with the increase or decrease of specimen size in metal-forming processes. The reduction of specimen size decreases the contacting area between the material and forming tool, and thus these intrinsic characteristics in the tool–workpiece interface, such as the open lubricate pocket and debris at the border region of contacting zone, become more dominating, which results in the change of overall friction force. Unlike asperity and feature sizes, the grain size effect is related to the influence of material properties on friction behaviors. The decrease of grain size may change the deformation or fracture of materials during the forming process, which further affects the overall friction resistance. Currently, there are no solid conclusions of grain size effect on friction, since different phenomena were observed in different experiments. In this section, the experimental observations of geometry and grain size effects on friction behaviors will be presented and their physical mechanisms will also be discussed.

3.2.1 Experimental Observation of Geometry Size Effect

The first approach to study the geometry size effect on friction was made by Messner [8] using ring compression tests, who found that an increase in friction causes a decrease in the specimen size. Such results were then confirmed using different experiments such as double-cup extrusion and cylinder compression tests.

3.2.1.1 Ring Compression

In ring compression test, friction condition at the tooling–workpiece interface can be evaluated based on the dimensional change of specimen. The flat rings with the dimensional proportion of 6:3:2 (outer diameter:inner diameter:height) are deformed. For a particular height reduction, the friction condition can be reflected by the change of inner diameter. The inner diameter change of the ring with different friction coefficients is shown in Figure 3.1.

To study the size effect on friction at the tooling–workpiece interface, Chan et al. [9] conducted a series of ring compression tests with different sizes of

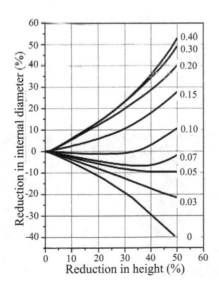

FIGURE 3.1
Calibration curves of friction coefficient in ring compression test [9].

ring specimens under the same lubricated condition. The original and compressed ring samples are shown in Figure 3.2, and the dimensions of the samples are presented in Table 3.1. As shown in Table 3.1, the measured friction coefficients in the three scenarios are 0.04, 0.05 and 0.075, respectively, which reveals that friction is increased with miniaturization.

Figure 3.3 shows the surface topography of the compressed samples. It is found that the asperities at the outer contacting area are flattened by the rigid tool, while the original shape of the asperities at the inner contacting area is mostly maintained. When the tooling presses the lubricated material

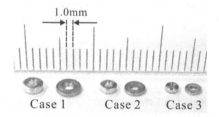

FIGURE 3.2
The original and compressed ring samples [9].

TABLE 3.1

Original Ring Dimensions, Change of Ring Dimensions, and the Estimated Friction Coefficients [9]

Case	Original Ring Dimensions (mm)			Final Inner Diameter (mm)	Final Height (mm)	Reduction of Inner Diameter (%)	Reduction of Height (%)	Friction Coefficient
	Outer Diameter	Inner Diameter	Height					
1	3	1.5	1	1.69	0.58	−12.92	41.61	0.04
2	2.5	1.25	0.83	1.36	0.48	−8.96	41.98	0.05
3	2	1	0.67	1.05	0.39	−5.20	41.53	0.075

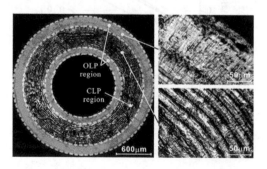

FIGURE 3.3
Surface topography of the specimen with convex peaks and concave valleys [9].

surface, the peaks of the convex features deform plastically. The lubricant could be trapped in the concave valleys or squeezed out. When the lubricant is squeezed out, the so-called open lubricant pocket (OLP) is thus formed. The asperities support the deformation load and are deformed to be flat. When the lubricant is trapped in the concave valleys, the so-called CLP is formed. Part of the deformation load is supported by the lubricant, resulting in the decrease of normal pressure on the asperities. The materials thus slide on the tooling surface with a lower friction.

3.2.1.2 Double-Cup Extrusion

Double-cup extrusion test, as shown in Figure 3.4, was proposed to evaluate the interfacial friction in extrusion process. It produces a relatively high surface expansion and high pressure and thus represents the actual situation in forward extrusion test [10]. A cylindrical billet is positioned in a die between a stationary punch and a moving one with an identical shape. When the upper punch moves downwards, the materials flow up and down, forming the upper and lower cups. If there is no friction, both the cups would have the same height, whereas at higher friction, the formation of the lower cup is prevented. This behavior resulted from the different relative velocities between moving and stationary punches, die and billet. Therefore, a small change in friction can be detected by a significant change in the heights of the lower and upper cups, which is usually expressed by the height ratio of the cups. Tiesler et al. [11] applied this test for investigations of the frictional size effect using geometrically similar specimens with an initial diameter of 4, 2, 1 and 0.5 mm and the ratio of diameter to height $D_0/H_0 = 1$.

The change in friction conditions with decreasing specimen size can be directly determined by comparing the cup height ratios, since the specimens are geometrically identical. To obtain the relationship of the friction factor m relating to a certain cup height ratio, FE simulation was widely used for the identification [12].

In simulation, three different relative punch strokes, viz., 35%, 50%, and 80%, were selected, and all the results are presented as the graphs of cup height ratio h_{upper}/h_{lower} versus relative punch stroke $z/H_0 \times 100\%$ (z is the

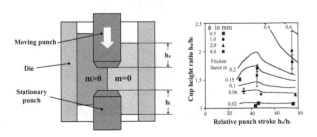

FIGURE 3.4
The formed geometries in different size-scaled double-cup extrusion tests [12].

punch stroke and H_0 is the initial specimen height). Figure 3.4 shows the cup height ratio versus the relative punch stroke of all the sample sizes, and the curves of constant friction determined by FE simulation. It can be clearly seen that friction increases with the decrease in specimen size. The numerical identification shows a friction factor of around $m = 0.02$ for the large specimens, and around $m = 0.06$ for the medium specimens, while it increases up to $m = 0.4$ for small specimens.

3.2.1.3 Cylinder Compression

The effect of specimen geometry size on interfacial friction was studied [10]. To avoid the effect of microstructural grain size on inhomogeneous deformation, the specimen geometry size effect was examined based on the compression of the specimen with the as-received fine microstructure. The compressed samples are shown in Figure 3.5.

Although the lubricant is applied on the tooling–workpiece interface, the friction on the contact surface cannot be totally eliminated. The friction prevents the material near the tooling surface from flowing outwards. This leads to the occurrence of the barreled side profile of the compressed specimen. Ebrahimi and Najafizadeh [14] conducted an analysis to estimate the friction factor (m) based on the upper bound theorem and the geometry of the testing specimen. The friction factor can be identified based on the following relation:

$$m = \frac{(R/H)b}{\left(4/\sqrt{3}\right)-\left(2b/\sqrt{3}\right)} \tag{3.1}$$

where R and H are the average radius and the height of the compressed specimen and b is the barreling parameter. R and b are given by

$$R = R_0\sqrt{\frac{H_0}{H}}$$

$$b = 4\frac{\Delta R}{R}\frac{H}{\Delta H} \tag{3.2}$$

FIGURE 3.5
The compressed specimens with different sizes: (a) Without lubrication; (b) Lubricated with castor oil [13].

where R_0 and H_0 are the initial radius and the height of specimen, respectively. ΔR is the difference between midheight radius (R_M) and top radius (R_T). ΔH is the compressed height. The top radius (R_T) of the compressed specimen is given by

$$R_T = \sqrt{3\frac{H_0}{H}R_0{}^2 - 2R_M{}^2} \tag{3.3}$$

The schematic illustration of the geometrical parameters is shown in Figure 3.6.

The friction factor can be determined based on Eqs. (3.1)–(3.3). It is found that the friction factor increases with the decrease of specimen size as shown in Figure 3.7.

Figure 3.8 shows the end surface topography of the compressed specimen. The light areas in Figure 3.9 are actually the real contact areas (RCAs), while the dark areas are the CLPs. The RCAs are concentrated at the outer rim of the end surface, while the CLPs are more located at the inner region of the workpiece.

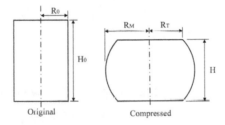

FIGURE 3.6
Geometrical parameters of the specimen [15].

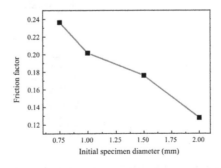

FIGURE 3.7
Friction factors in compressing pure copper cylinders with different diameters [15].

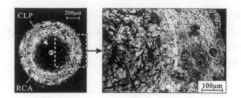

FIGURE 3.8
End surface topography of the compressed specimen with the original dimensions of $\varnothing 0.7 \times 1.125\,mm$ [15].

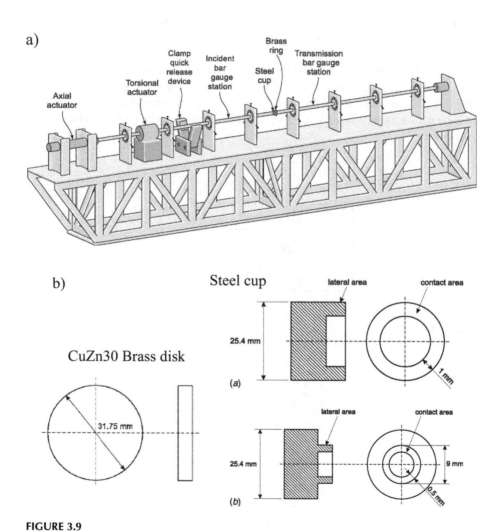

FIGURE 3.9
(a) Scheme of the stored-energy Kolsky bar and (b) specimens used for the dynamic friction analysis [16].

3.2.2 Experimental Observations of Grain Size Effect

As the flow behavior of materials is significantly affected by the grain size, the indirect friction tests, such as ring and cylinder compression and double-cup extrusion, are considered invalid in studying the grain size effect on friction and wear behaviors of ductile materials. Experimental observations in this field are thus usually conducted using direct friction tests such as sphere/pin-on-disk friction, Kolsky bar friction test, and block-on-ring friction test. Recently, Peng et al. [4] conducted a friction test of different grain-sized pure copper pin against steel disk. In their experiments, two types of friction processes, namely, adhesive and plowing frictions, were designed based on the theory of Wilson [8], and the grain size effects on these two types of friction process were studied separately. The adhesive friction is referred to as the deformable rough copper (RC) surface sliding on a smooth-ridged steel surface, in which the friction stress is considered to be dominated by adhesion. The plowing friction, on the other hand, is defined as the rigid rough steel (RS) surface scratching the deformable smooth copper (SC) surface, in which plowing is considered as the dominating mechanism.

3.2.2.1 Kolsky Bar Friction Test

A series of frictional experiments was conducted by Mori et al. [16] using a stored-energy Kolsky bar. The scheme of the stored-energy Kolsky bar used for the dynamic friction analysis is shown in Figure 3.9. It contains 25.4-mm aluminum alloy (7075-T6) bars; the incident bar is 2.3 m long and the so-called transmission bar is 1.9 m long. The compression, tension, and shear loading are produced by hydraulic actuators: an axial hydraulic double-acting actuator applies a compressive or tensile load at one end of the incident bar, and a hydraulic rotary actuator located along the incident bar applies the torque. The axial load is applied before gripping the clamp, and the friction phenomenon is studied under quasistatic pressure conditions. On release of the clamp, a torsional pulse, with a constant amplitude equal to one-half of the stored torque, propagates down the bar toward the specimen. Simultaneously, an unloading pulse of equal magnitude propagates from the clamp toward the rotary and axial actuators. As the pulse travels down the bar, it is detected by two strain gage stations: one on the incident bar and the other on the transmission bar.

The Kolsky bar friction test method is used to directly measure the existing interfacial conditions, which does not depend on material deformation behavior like other methods to measure friction. The method also provides both static and dynamic coefficients of friction. The tests of Mori et al. [16] were conducted using brass samples with a small grain size (32 μm) and a large grain size (211 μm), and the low contact pressure (22 MPa) and high contact pressure (250 MPa) to see the change of friction due to variation of these parameters.

As shown in Figure 3.10, the specimens with a larger grain size have a higher average friction coefficient in both static and dynamic scenarios measured in experiments. However, due to the large scatter of the measured value, the authors did not conclude a significant effect of grain size on friction.

3.2.2.2 Block-on-Ring Friction Test

The effect of grain size on the friction and wear behaviors of copper (Cu) samples under different lubricant conditions was studied by Moshicovich et al. [17]. All the friction tests were conducted under laboratory conditions using a block-on-ring rig, as shown in Figure 3.11.

The Cu samples with different grain sizes (1, 30, and 60 μm) were used in the experiments, as shown in Figure 3.12. Particularly, the Cu samples with the average grain size of 1 μm were obtained by equal channel angular pressing for the severe plastic deformation of material.

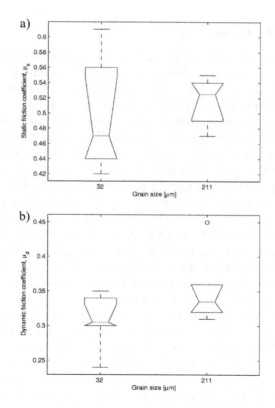

FIGURE 3.10
Results of Mori's experiments: grain size effect on (a) static and (b) dynamic coefficient of friction [16].

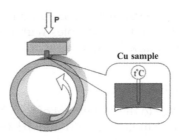

FIGURE 3.11
Block-on-ring test rig [18].

FIGURE 3.12
Microstructure of the Cu samples with different grain sizes. (a) 1 μm, (b) 30 μm, (c) 60 μm [17].

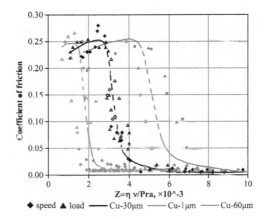

FIGURE 3.13
The Stribeck curves for the copper–steel pairs with different grain sizes of Cu. The circles indicate the effect of sliding velocity on the friction coefficient. The triangles indicate the effect of load on the friction coefficient. The black marks characterize the steady friction state, whereas the blank marks characterize the unsteady range [17].

The Stribeck curves for Cu samples with different virgin grain sizes were considered. Elastohydrodynamic lubrication (EHL) and boundary lubrication (BL) regions were mainly studied with the results shown in Figure 3.13. Similar Stribeck curves were found out for Cu samples with different virgin grain sizes. A load of the transition from the EHL to BL region was increased with a decrease of the grain size. While the friction coefficients were similar in the EHL and BL regions for the samples with different grain sizes, the wear rate was increased remarkably with an increase of the virgin grain size.

3.2.2.3 Pin-on-Disk Friction Test

Friction force is a resistance to the movement of tool. The real friction process is a sum of different friction models. In different friction modes, the source of resistance is different. For example, the resistance in adhesion mode comes from the shear of welding junctions between tooling and workpiece. In plowing mode, however, the resistance is caused by the deformation of the workpiece. The grain size effects on friction in these two friction modes are thus believed to be different from each other. Peng et al. [4] conducted a series of pin-on-disk friction tests to study the grain size effect in adhesive and plowing friction modes. In their experiments, the friction pair was designed to be composed of a copper pillar and a steel disk. The geometries of the friction pair are shown Figure 3.14. Two kinds of surface pairs, viz., SC–RS pair and RC–smooth steel (SS), were designed to study the adhesive and plowing frictions separately.

In addition, different grain-sized copper pillars were prepared by vacuum heat treatment to investigate the grain size effect on friction behavior, as shown in Figure 3.15.

To reduce the influence of the contact area caused by normal load, the testing loading is normalized by yield stress as

$$F^* = \frac{F}{\sigma_0 A} \tag{3.4}$$

where F^* is the normalized normal load. F is the normal load, σ_0 is the initial yield stress of copper, and A is the nominal contact area. The grain size effect on friction is measured under the same normalized normal load. Two pairs of contacting surfaces were assembled to investigate the adhesive and plowing frictions, respectively. The design of experiments is shown in Table 3.2.

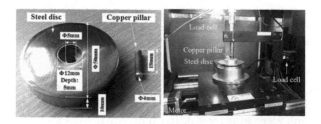

FIGURE 3.14
Geometries of the testing samples [4].

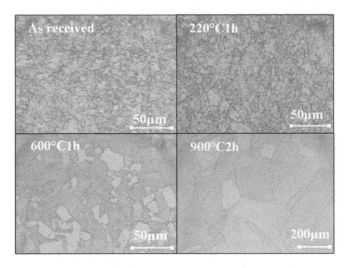

FIGURE 3.15
Microstructures of the copper samples prepared by different heat treatment processes [4].

TABLE 3.2

Design of the Friction Tests [4]

Process	Steel Surface Ra (μm)	Copper Surface Ra (μm)	Copper Grain Size (μm)	Normalized Normal Loads (F^*)
Adhesive friction	0.05	3.2 ± 0.18	4.3 ± 0.18	0.03, 0.05, 0.07
		3.1 ± 0.02	11.4 ± 1.3	
		2.8 ± 0.66	132.2 ± 17.7	
Plowing friction	1.74	0.027 ± 0.007	4.3 ± 0.18	0.005, 0.008, 0.011
		0.026 ± 0.008	11.4 ± 1.3	
		0.025 ± 0.004	132.2 ± 17.7	

The friction coefficient μ defined as a function of time t for different grain sizes and normalized normal load is presented in Figures 3.16 and 3.17. The friction coefficient is determined based on the normal and tangential loads measured by load cells. For the adhesive friction tests, the friction coefficient, as shown in Figure 3.16, is first increased from 0 to a peak value, and then decreased to a stable value rapidly. The peak value is the static friction coefficient and the stable value is the dynamic one.

For the plowing friction test, the shape of friction coefficient–test time curve is more complicated. A nearly linear increasing curve followed by an oscillating curve was observed for most samples, as shown in Figure 3.17. The linear part is considered to be the static friction, while the oscillating part represents the friction coefficient when the steel disk rotates. The maximum static friction coefficient is the linear limit of the curves. It is probably smaller than the dynamic coefficient in plowing friction test, since the

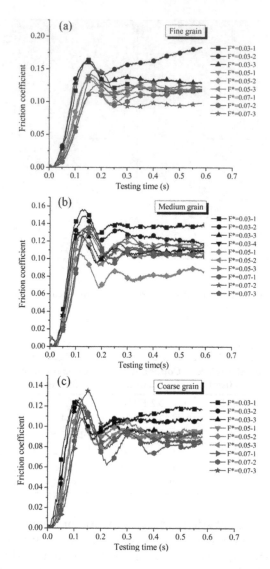

FIGURE 3.16
Variation of friction coefficient with time in the adhesive friction test with different normal pressures and grain sizes [4].

pileup of material increases the contact area of the steel asperity and the copper surface during the process.

The maximum static friction coefficient, μ, measured in adhesion test under the normalized normal load of 0.03, 0.05, and 0.07 for the samples with different grain sizes is given in Figure 3.18a. It is clear that the friction coefficient decreases with the grain size and the normalized normal load. The results obtained when $F^* = 0.07$, on the other hand, do not match the grain size effect

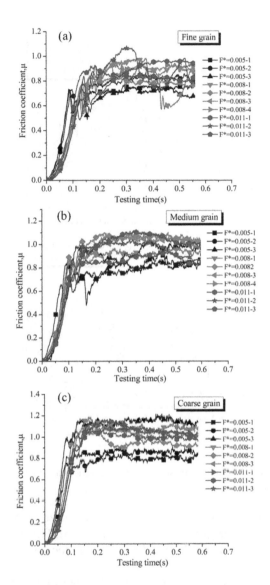

FIGURE 3.17
Variation of friction coefficient with time in the plowing friction test with different normal pressures and grain sizes [4].

occurring under other normal loads. The medium-grained samples possess a larger friction coefficient than the fine-grained samples under the normalized normal load of 0.07. Moreover, for the samples with medium and coarse grains, the friction coefficient under the normalized normal load of 0.07 is higher than that obtained under the normalized normal load of 0.05. These deregulations may be attributed to the effect of the tiny asperities on the SS surface plowing the deformed RC surfaces. The obtained roughness of

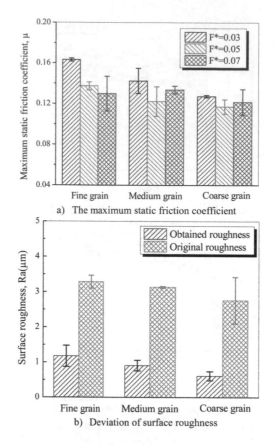

a) The maximum static friction coefficient

b) Deviation of surface roughness

FIGURE 3.18
Effect of grain size on the maximum static friction coefficient and the deviation of surface roughness in adhesive friction tests [4].

the samples after adhesion test under the normalized normal load of 0.05 is compared with the original roughness, shown in Figure 3.18b. It is found that the average value of the obtained roughness decreases with the increase of grain size.

The grain size effect on the maximum static friction coefficient of plowing tests is shown in Figure 3.19. Opposite to the observation in adhesive friction test, the samples with a larger grain size have the higher average value of μ under all normal loads in plowing friction tests. No significant relation is found between the grain size effect and the normalized normal load. The surface roughness after plowing test of $F^* = 0.008$ is presented in Figure 3.19a. Compared with the original roughness, an increase of roughness with more than 0.6 μm is obtained after the test. The roughening of the surface is mainly caused by the steel asperity plowing the copper surface. An increasing tendency of the roughness with grain size is observed. As the original roughness of the samples has only a little difference of 0.002 μm, the increased

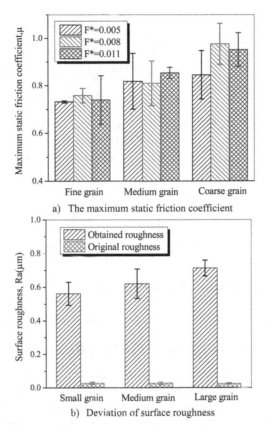

FIGURE 3.19
Effect of grain size on the maximum static friction coefficient and the deviation of surface roughness in plowing friction test [4].

difference between the surface roughness of the samples with different grain sizes is attributed to the grain size effect.

The grain size effect on the surface morphology after the plowing test is presented in Figure 3.20. The plowing tracks and damages on the copper surfaces shown in Figure 3.20b–d are caused by the asperities on the steel surface in Figure 3.20a. By comparing Figure 3.20b–d, a significant difference is easily found between the plowing tracks of the samples with fine and coarse grains. For the samples with fine grains, the edges of the tracks appear to be blurry. For the samples with coarse grains, however, the much clearer and more continuous edges of the tracks are formed. Since the material damage is significant in plowing process, a transformation of the damage mechanism is believed to be the reason for the difference in the surface morphology. A similar transformation mechanism was reported by Senda et al. [19], where transgranular cracks were found in the wear surface of the large-grained specimens and the intergranular cracks in the fine-grained samples.

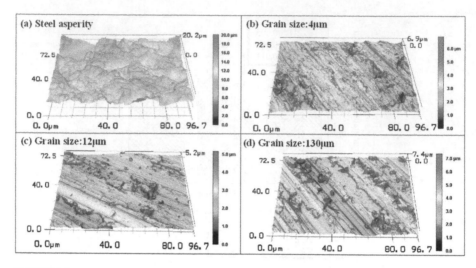

FIGURE 3.20
Comparison of the surface topography after plowing test [4].

3.3 Modeling and Analysis of Friction Behavior

In this section, the analysis of mechanisms of both the grain and feature size effects on friction behaviors is presented. In modeling of grain size effect, the adhesive friction and plowing friction are treated separately, based on the significant difference in surface evolution observed in experiments.

3.3.1 Grain Size Affected Adhesive Friction Behavior

According to the descriptions in Section 3.2.2.3, when the roughness of rigid tool is extremely small, the overall friction force can be considered to be the adhesive force on the RCA in tool–workpiece interface. Since the surface of workpiece is not smooth, the RCA is formed by the deformation of surface asperities of workpiece, which is affected by the mechanical property of materials such as yield stress and strain hardening. Small welding junctions are then generated between the deformed asperities and the surface of the tool, and the adhesive stress equals to the shear of the junctions once sliding initiates. According to Tabor's study on adhesion [20], the relative sliding can cause an increase of RCA, which is called "junction growth," due to the plastic flow of material. The increase of RCA in turn increases the overall friction force. To model the adhesive friction, the influence of material hardening and junction growth should be considered. The grain size effect on adhesive friction is revealed through grain size dependency of mechanical properties.

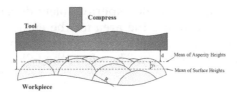

FIGURE 3.21
Contact model of nominally flat rough surface [22].

According to the classical contact mechanics, the contact between the tool and workpiece is equivalently represented by a single rough surface and a rigid flat, as schematically described in Figure 3.21 [21].

In the figure, two reference planes, the mean of asperity heights and the mean of surface heights, are represented by two dashed lines in Figure 3.21. The distance between the rigid flat and the mean of asperity heights is denoted by separation d, while the distance between the rigid flat and the mean of surface heights is designated as h. Let y_s represent the separation of the two reference planes, and z the height of the asperity measured from the mean of asperity heights, as shown in Figure 3.21.

When the rigid flat tool contacts the deformable rough workpiece, the interference of an asperity is dependent on its height. For an asperity with the height of z, the interference can be defined as

$$\omega = z - d \tag{3.5}$$

Gaussian distribution is employed in this model to describe the height of asperity and is given by

$$\phi(z) = \frac{1}{\sqrt{2\pi}\sigma_s} \exp\left(-0.5\left(\frac{z}{\sigma_s}\right)^2\right) \tag{3.6}$$

where σ_s is the standard deviation of the asperity heights.

The following assumptions are made in modeling the contact process of asperities:

1. Bulk deformation is neglected during the contact and friction processes.
2. No interaction among asperities is considered.
3. All the asperities are assumed to be spherical, at least near their summits with a uniform radius R, and the height varies randomly.

Based on assumption 2, the normal load, P, the tangential load at sliding inception, T, the contact areas due to the normal load alone, A_0, and the area at sliding inception, A_i, of a single asperity, solely depend on its interference, ω. Therefore, the load and the area for the entire rough surface at a given separation d can be calculated by summing up the loads and area of each contact asperity, respectively. Therefore, the following equations can be obtained:

$$F_n = \eta A_n \int_d^\infty P_{\text{asperity}}(z-d)\phi(z)dz \tag{3.7}$$

$$F_t = \eta A_n \int_d^\infty T_{\text{asperity}}(z-d)\phi(z)dz \tag{3.8}$$

$$S_0 = \eta S_n \int_d^\infty A_0(z-d)\phi(z)dz \tag{3.9}$$

$$S_i = \eta S_n \int_d^\infty A_i(z-d)\phi(z)dz \tag{3.10}$$

where η and A_n are the asperity density and the nominal area of rough surfaces, respectively.

To model the grain size effect on adhesive friction of rough surfaces, the asperity friction model considering the influence of the mechanical properties of material should be established first. By summing up the behaviors of asperities in the rough surface based on Eqs. (3.7)–(3.10), the friction model of the rough surface considering mechanical properties of material is obtained. Employing the grain size effect on mechanical properties of material, the grain size effect on adhesive friction is revealed.

3.3.1.1 Asperity Contact Model

According to Greenwood and Williamson [23], the contact problem could be simplified to a rigid flat plane compressing a deformable hemisphere, as shown in Figure 3.22.

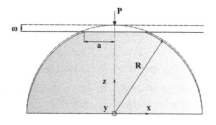

FIGURE 3.22
A deformable hemisphere pressed by a rigid flat plane [22].

In Figure 3.22, the solid and dashed lines show the situations after and before the deformation. In addition, three kinds of deformations, viz., fully elastic, elastic–plastic, and fully plastic, can be observed with the increase of the normal load.

In fully elastic deformation, the mean pressure, P_{ae}, and the contact area, A_e, of the contact surface, are given by [24]

$$p_{ae} = \frac{4E_H}{3\pi}\left(\frac{\omega}{R}\right)^{\frac{1}{2}} \tag{3.11}$$

$$A_e = \pi R \omega \tag{3.12}$$

E_H is the Hertz elastic modulus defined as

$$\frac{1}{E_H} = \frac{1-v_1^2}{E_1} + \frac{1-v_2^2}{E_2} \tag{3.13}$$

where E_1, E_2 and v_1, v_2 are Young's modulus and Poisson's ratios of the two materials. If one of the mating materials is much stronger than the other, Eq. (3.13) can be simplified as

$$\frac{1}{E_H} = \frac{1-v^2}{E} \tag{3.14}$$

where E and v are the elastic modulus and the Poisson's ratio of the softer material, respectively.

According to Chang et al. [25], the critical interference, ω_1, which marks the transition from elastic to elastic–plastic contact, is formulated as

$$\omega_1 = \left(\frac{\pi K_m H}{2E_H}\right)^2 R \tag{3.15}$$

where K_m is a constant, with relation to the Poisson ratio of the sphere by $K_m = 0.454 + 0.41v$. H is the hardness of the sphere, relating to its yielding strength Y by $H = K_h Y$, for elastic perfectly plastic materials. K_h is a constant with a predicted value of approximately 3 according to Tabor. For materials with a hardening character, the yield stress is termed as flow stress when yield happens [26,27].

Jackson and Green [28] claimed that the hardness is not a material constant as suggested by Tabor. According to their study, with the increase of interference and the change of contact geometry, the ratio of the limiting average pressure to yielding strength must change from Tabor's predicted value of 3 to a theoretical value of 1. An empirical formation of K_h is thus given as

$$K_h = 2.84\left[1 - \exp\left(-0.82\left(\frac{a}{R}\right)^{-0.7}\right)\right]$$

(3.16)

where a is the radius of the contact surface.

In fully plastic deformation, the average contact pressure equals to the hardness of material, regardless of the increase of normal interference according to Zhao et al. [24]. The contact area A_p and the mean contact pressure P_{pa} are given by

$$A_p = 2\pi R\omega$$

(3.17)

$$p_{pa} = H$$

(3.18)

Zhao et al. [24] proposed a contact model for elastic–plastic period by making a smooth transition from fully elastic to fully plastic period. The elastic–plastic deformation in the model is represented as

$$A_{ep} = \pi R\omega\left[1 - 2\left(\frac{\omega - \omega_1}{\omega_2 - \omega_1}\right)^3 + 3\left(\frac{\omega - \omega_1}{\omega_2 - \omega_1}\right)^2\right]$$

(3.19)

$$p_{epa} = H\left[1 - (1 - K_c)\frac{\ln\omega_2 - \ln\omega}{\ln\omega_2 - \ln\omega_1}\right]$$

(3.20)

where A_{ep} and P_{epa} are the area and the average pressure of the contact surface, respectively. $K_c = \frac{2}{3}K_m$ is a constant for an average contact parameter. ω_1 and ω_2 are the critical interferences corresponding to the critical normal pressure. The relation of ω_1 and ω_2 is designated as $\omega_2 = 54\,\omega_1$ for perfectly plastic materials. Furthermore, Kogut et al. [29] proposed a different coefficient of 110 for the relation in his study of the transition from elastic to plastic in detail, with an assumption of perfectly plastic material using FE simulation. An elastic core was also found in the elastic–plastic period.

3.3.1.2 Modified Sphere-Flat Contact Model Considering Material Hardening Effect

Mao et al. [22] extended the contact model to be able to handle and consider the effect of material hardening via addressing two issues: (1) The material hardness, denoted as H, varies with the strain on contact surface, and the strain is determined by interference and material properties and (2) The relation of the two critical interferences, ω_1 and ω_2, changes with the varying hardness. In Mao's model, the material of the sphere follows a power hardening law in the following:

$$\sigma = \begin{cases} E\varepsilon, & \varepsilon \leq \varepsilon_0 \\ C\varepsilon^n, & \varepsilon > \varepsilon_0 \end{cases} \tag{3.21}$$

where E is the Young's modulus, C is a material constant, and n is the strain-hardening exponent. Strain hardening is accommodated as soon as the plastic deformation happens, i.e., $\omega > \omega_1$.

The expressions of normal pressure and contact area can thus be derived as

$\omega \leq \omega_1$:

$$p_{ae} = \frac{4E}{3\pi} \left(\frac{\omega}{R} \right)^{\frac{1}{2}}$$

$$A_e = \pi R \omega$$

$\omega_1 < \omega \leq \omega_2^{*}$:

$$p_{aep} = K_T K_H C \left(\frac{2}{\sqrt{3}} K_w \left(\frac{\omega}{R} \right)^{\frac{1}{2}} \right)^n \tag{3.22}$$

$$A_e = K_S \pi R \omega$$

$\omega > \omega_2^{*}$:

$$p_{ap} = K_H C \left(\frac{2}{\sqrt{3}} K_w \left(\frac{\omega}{R} \right)^{\frac{1}{2}} \right)^n \tag{3.23}$$

$$A_p = 2\pi R \omega$$

The critical interference, ω_2^{*}, is given by

$$\omega_2^{*} = \left[\frac{200 K_c Y \left(\frac{\pi K_m K_H Y}{2E} \right)^2}{C \left(\frac{2}{\sqrt{3}} K_w K_p \right)^n} \right]^{\frac{1}{1+\frac{1}{2}n}} R \tag{3.24}$$

K_T, K_S, and K_P are the parameters given as follows:

$$
K_T = \begin{cases} 1-(1-K_c)\dfrac{\ln\omega_2^{*}-\ln\omega}{\ln\omega_2^{*}-\ln\omega_1}, & \omega \in \left[\omega_1,\omega_2^{*}\right] \\[3mm] 1, & \omega \in \left[\omega_2^{*},\infty\right) \end{cases}
$$

$$
K_S = \begin{cases} 1-2\left(\dfrac{\omega-\omega_1}{\omega_2^{*}-\omega_1}\right)^3 + 3\left(\dfrac{\omega-\omega_1}{\omega_2^{*}-\omega_1}\right)^2, & \omega \in \left[\omega_1,\omega_2^{*}\right] \\[3mm] 2, & \omega \in \left[\omega_2^{*},\infty\right) \end{cases} \tag{3.25}
$$

$$
K_w = \frac{2}{n+1}\sqrt{\frac{K_1}{\sqrt{3}\pi K_T K_S K_H (n+1)}}
$$

$$
K_p = \frac{\varepsilon_0}{\dfrac{2}{\sqrt{3}}\dfrac{2}{n+1}\sqrt{\dfrac{K_1}{\sqrt{3}\pi K_H (n+1)}}\left(\dfrac{\omega_1}{R}\right)^{\frac{1}{2}}}
$$

3.3.1.3 Interfacial Friction Stress and Junction Growth

According to Johnson et al. [30] and Wanheim et al. [31], the friction stress on the deformed surface asperity in metal working process is assumed to be constant with the formulation as follows:

$$
\tau = mK \tag{3.26}
$$

where m is a positive parameter defined as the friction factor with a value between 0 and 1, and K is the shear strength of the soft material of the contact couple.

Tabor et al. [32] claimed that junctions were distorted before shearing and resulted in an increase of the contact area. Tabor's equation to describe the growth of junction is

$$
p^2 + \alpha\tau^2 = H^2 \tag{3.27}
$$

where τ and p are the tangential stress and the normal pressure of the contact surface, and H is the hardness of the soft material. The constant α, depending on the stress state of the junction, can be determined by assuming the normal pressure as zero and formulated as

$$
\alpha = \frac{H^2}{K^2} = 3K_h^{2} \tag{3.28}
$$

According to Chaudhri [33], no significant junction growth would happen if $p^2 + a\tau^2 \leq H^2$. Junction growth starts when $p^2 + a\tau^2 > H^2$, and the area growth thus leads to the decrease of normal pressure until Eq. (3.27) is satisfied again. Assuming that the contact areas at the beginning and end of the junction growth are A_0 and A_{01}, respectively, they must satisfy the following equation.

$$p_0 A_0 = p_{01} A_{01}$$

Hence, the ratio of junction growth can be represented as

$$\gamma = \frac{A_{01}}{A_0} = \frac{p_0}{p_{01}} \qquad (3.29)$$

Considering Eq. (3.26), the normal pressure after junction growth is obtained:

$$p_{01} = \sqrt{H^2 - \alpha(fK)^2} = H\sqrt{1 - \alpha\left(\frac{m}{\sqrt{3}K_h}\right)^2} \qquad (3.30)$$

Equation (3.30) can be used as the critical pressure to determine the occurrence of junction growth. And Eq. (3.29) can be further formulated as

$$\gamma = \frac{p_0}{H\sqrt{1 - \alpha\left(\frac{m}{\sqrt{3}K_h}\right)^2}} \qquad (3.31)$$

If $p_0 \leq p_{01}$, junction growth cannot initiate and the following conditions are satisfied:

$$\tau = mK, \quad p = p_0, \quad \gamma - 1$$

Otherwise, if $p_0 > p_{01}$, junction growth should be considered. The friction stress and junction growth are designated as

$$\tau = mK, \quad p = H\sqrt{1 - \alpha\left(\frac{m}{\sqrt{3}K_h}\right)^2}, \quad \gamma = \frac{p_0}{H\sqrt{1 - \alpha\left(\frac{m}{\sqrt{3}K_h}\right)^2}}$$

3.3.1.4 Asperity Friction Model

The critical pressure that characterizes the occurrence of junction growth p_{01} corresponds to a critical interference ω_b. As K_h approximates to 2.8 for the small interferences from ω_1 to ω_2^*, the value of ω_b depends on the friction factor m. By substituting Eq. (3.30) into Eq. (3.22), the formulation of ω_b can be obtained as

$$\omega_b = \omega_2^* \left(\frac{\omega_1}{\omega_2^*} \right)^{\frac{1-\sqrt{1-\alpha\left(\frac{m}{\sqrt{3}K_h}\right)^2}}{1-K_c}} \tag{3.32}$$

The whole deformation of the asperity due to the increase of the interference can be divided into four stages based on three critical interferences: ω_1, ω_b, and ω_2^*.

Stage 1: $\omega \leq \omega_1$, fully elastic contact, no effect of hardening or junction growth.

$$p_1 = \frac{4E}{3\pi} \left(\frac{\omega}{R} \right)^{\frac{1}{2}} \tag{3.33}$$

$$A_1 = \pi R \omega \tag{3.34}$$

$$\tau_1 = mK \tag{3.35}$$

Stage 2: $\omega_1 < \omega \leq \omega_b$, elastic–plastic contact with material hardening, and no effect of junction growth

$$p_2 = K_T K_H C \left(\frac{2}{\sqrt{3}} K_w \left(\frac{\omega}{R} \right)^{\frac{1}{2}} \right)^n \tag{3.36}$$

$$A_2 = K_S \pi R \omega \tag{3.37}$$

$$\tau_2 = \frac{mC}{\sqrt{3}} \left(\frac{2}{\sqrt{3}} K_w \left(\frac{\omega}{R} \right)^{\frac{1}{2}} \right)^n \tag{3.38}$$

Stage 3: $\omega_b < \omega \leq \omega_2^*$, elastic–plastic contact with material hardening and junction growth.

$$p_3 = K_H C \left(\frac{2}{\sqrt{3}} K_w \left(\frac{\omega}{R} \right)^{\frac{1}{2}} \right)^n \sqrt{1 - \alpha \left(\frac{m}{\sqrt{3}K_h} \right)^2} \tag{3.39}$$

$$A_3 = \gamma K_S \pi R \omega = \frac{K_T K_S}{\sqrt{1 - \frac{\alpha}{3} \left(\frac{m}{K_h} \right)^2}} \pi R \omega \tag{3.40}$$

$$\tau_3 = \frac{1}{\sqrt{3}} C \left(\frac{2}{\sqrt{3}} K_w \left(\frac{\omega}{R} \right)^{\frac{1}{2}} \right)^n \tag{3.41}$$

Stage 4: $\omega > \omega_2^*$, fully plastic contact with material hardening and junction growth.

$$p_4 = K_H C \left(\frac{2}{\sqrt{3}} K_w \left(\frac{\omega}{R} \right)^{\frac{1}{2}} \right)^n \sqrt{1 - \alpha \left(\frac{m}{\sqrt{3} K_h} \right)^2} \tag{3.42}$$

$$A_4 = \frac{2}{\sqrt{1 - \frac{\alpha}{3} \left(\frac{m}{K_h} \right)^2}} \pi R \omega \tag{3.43}$$

$$\tau_4 = \frac{1}{\sqrt{3}} C \left(\frac{2}{\sqrt{3}} K_w \left(\frac{\omega}{R} \right)^{\frac{1}{2}} \right)^n \tag{3.44}$$

3.3.1.5 Grain Size Effects

According to Mao's model, the adhesive friction behavior is affected by the mechanical properties of the material, especially plasticity. As is known, the flow stress of material is significantly affected by the grain size. Thus, the change of grain size may change the adhesive friction behaviors. To determine the mechanism of grain size effect on adhesive friction behaviors, Mao's model is employed to predict the experimental observations in Section 3.2.2.3.

In adhesive friction, the steel surface could be simplified to a rigid flat plane, and the copper could be represented by a deformable rough surface, depending on the difference of the hardness and the surface roughness between the tool and the workpiece [34]. When the normal load is applied, the asperities on the workpiece surfaces are deformed by the rigid tool. In fact, both elastic and plastic deformations exist on the workpiece surface since the height of asperity is not uniform. To identify whether the surface is likely to deform plastically or not, the plasticity index, Ψ, was proposed [23,25].

$$\Psi = \frac{4E}{3\pi K_c H} \left(\frac{\sigma}{R} \right)^{\frac{1}{2}} \left(1 - \frac{3.717 \times 10^{-4}}{(\eta \sigma R)^2} \right)^{\frac{1}{4}} \tag{3.45}$$

where K_c, E, and H are material-related parameters, representing the maximum contact pressure factor, Hertz elastic modulus, and the hardness of softer material in the contacting mates, respectively. The formulations of K_c and E are given by

$$K_c = 0.454 + 0.41v \tag{3.46}$$

$$E = \frac{E_s}{1 - v^2} \tag{3.47}$$

where E_s and v are the elastic modulus and Poisson's ratio of the softer material. R and η in Eq. (3.45) are the roughness-related parameters representing the average radius and the density of the asperity summits. The detailed representations are given by [35]

$$R = \frac{3\sqrt{\pi}}{\sigma''} \tag{3.48}$$

$$\eta = \frac{(\sigma''/\sigma')^2}{6\pi\sqrt{3}} \tag{3.49}$$

where σ, σ', and σ'' are the root mean square (*RMS*) height, *RMS* slope, and *RMS* curvature of the surface. All these parameters can be obtained based on the surface profile $y(x)$ [25], which can be measured by a digital laser microscope. In this experiment, the values of the dimensionless parameters are obtained as $\eta\sigma R = 0.223$, $\sigma/R = 4 \times 10^{-4}$. Assuming that the surfaces of copper samples are similar, the plasticity indexes of the copper samples with different grain sizes are obtained by substituting the surface parameters and mechanical parameters into Eq. (3.45). The results are shown in Figure 3.23.

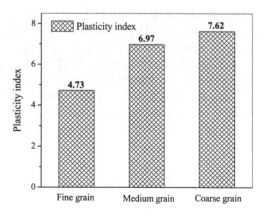

FIGURE 3.23
Plasticity index of specimens [4].

It is obvious that the plasticity index increases with grain size, since the E/H ratio of the sample increases with grain size.

According to Kogut and Etsion [36], for the surfaces with $0.6 < \psi < 8$, both elastic and plastic deformations occur during the contact. The friction coefficient decreases with the increase in plasticity index [37–40]. Assuming the hardening exponential $n = 0$ and substituting the surface and material parameters into Mao's model, the friction coefficient as a function of the normalized normal load is calculated and compared with the experimental results, as shown in Figure 3.24.

According to Figure 3.24, the predicted results reveal the same tendency as the experimental ones. The friction coefficient decreases with the increase of both grain size and the normalized normal load. Although a difference is observed between the predicted values and the measured ones, the predicted error is acceptable since most of the predicted values are located within the error bar. And the discrepancy of the model and experiment may be caused by the measurement error of the test facility.

The theoretical model reveals that the grain size effect on the coefficient value of adhesive friction is attributed to the fact that samples with a larger grain size have a higher E/H ratio, which causes a higher value of plasticity index of the contact surface and makes the deformation more plastic. The friction coefficient thus decreases with grain size. When the normal load increases, all the samples tend to deform plastically and the discrepancy among the different grain-sized samples gets smaller. This is why the grain size effect decreases with the normal load. On the other hand, the plasticity of the surface also affects the ending surface. Since the samples with a lower plasticity index have more elastic recovery and the surfaces are flattened by a rigid tool, the recovery after unloading is thus beneficial to the roughening of ending surfaces.

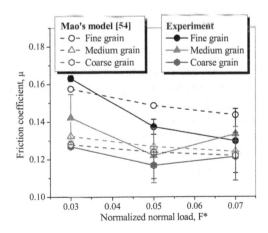

FIGURE 3.24
Comparison of the friction coefficient predicted by Mao's model with the experiment [4].

3.3.2 Grain Size Affected Plowing Friction Behavior

To identify the mechanism of grain size effect on plowing friction, the theoretical analysis based on elastic–plastic deformation and fracture mechanics is presented in this section.

3.3.2.1 Plowing Model Considering Elastic Recovery

Based on the classical friction model, the friction force resulted from the force of steel asperity plowing the copper surface in the plowing process. According to contact mechanics, the steel asperity is assumed to be a rigid conical indenter and the surface of the softer material is assumed to be deformably flat. The friction coefficient for an ideal rigid-plastic material is given as [41]

$$\mu_p = \frac{2}{\pi}\cot\theta \tag{3.50}$$

where θ is the semiangle of the indenter. According to Eq. (3.50), the plowing friction coefficient depends only on the surface profiles of the rigid surface.

Recent studies have shown that the mechanical property of the deformable material is another important factor that affects the friction coefficient. Bressan et al. [42] found that the coefficient of friction decreases when elasticity becomes more predominant and becomes zero when elastic deformation is the only deformation. Bucaille et al. [43] attributed this phenomenon to the elastic recovery, and adopted the rheological factor to measure the ratio between the deformation imposed by indenter and the part of elastic deformation. The rheological factor is given by [43]

$$X = \frac{E}{\sigma_0}\cot\theta \tag{3.51}$$

When X is close to 10, the deformation is mainly elastic; and if $X > 100$, the deformation is mainly plastic. According to Bucaille et al., for the small value of X [39], elastic recovery is large at the rear and on the side of the indenter. The pressure on the rear side of indenter offsets the resistance force on the front side. Thus, the smaller X has a smaller friction coefficient. An angular parameter α was introduced to model the rear contact area, and the friction model was improved as [43]

$$\mu = \frac{\sin\left(\alpha + \dfrac{\pi}{2}\right)}{\alpha + \dfrac{\pi}{2}}\cot\theta \tag{3.52}$$

Obviously, α is a function of X, and the numerical relation is presented in Figure 3.25.

Based on the data points shown in Figure 3.25 and the fact that $\alpha = 0$ when $X = \infty$, the function is fitted as follows:

$$\alpha = -0.55232 + \frac{191.50164}{(1 + 2.93574X)^{\frac{1}{1.53876}}} \tag{3.53}$$

Actually, this function is unavailable if $X < 1$, because another boundary condition ($\alpha = 90$, if $X = 0$) is not satisfied.

3.3.2.2 Plowing Friction Behaviors Considering Fracture Behaviors

As material damage and abrasion are observed on the surface of the specimens, as shown in Figure 3.20, the fracture behaviors of material are considered to be another important reason for grain size effect on plowing behavior.

As mentioned earlier, two modes of fracture, viz., transgranular and intergranular fractures [44], exist in friction process. When the indented depth is smaller than the grain size, plowing only happens within the surface grains. The plowing force mainly comes from the transgranular fracture of surface grains. However, if the indented depth is much larger than the grain size, the intergranular fracture becomes more dominating than the transgranular fracture, since it is easier for microvoids to generate and propagate at grain boundaries than the inner part of grains [45]; therefore, the stress needed to cause intergranular is smaller [46].

Figure 3.26 shows the mechanism of grain size effect on plowing friction for polycrystalline metallic materials. With a given normalized normal load,

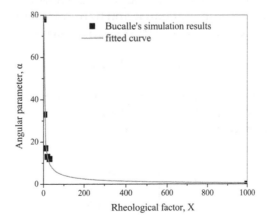

FIGURE 3.25
Angular parameter as a function of the rheological factor [4].

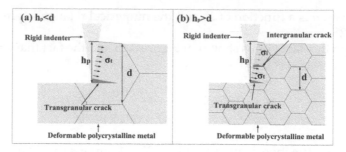

FIGURE 3.26
Grain size effect on plowing friction of polycrystalline metallic materials [4].

the indented depth, h_p, of the samples is almost the same and its average value is around 6.5 μm according to Figure 3.20b–d. For the samples with coarse grains, the grain size ($d = 132$ μm) is much larger than the indented depth. Thus, damage only happens on the surface grains, and the fracture mode should be a transgranular fracture. For the samples with fine grains, as the grain size ($d = 4$ μm) is smaller than the indented depth, intergranular fracture dominates the plowing process. For the samples with medium-sized grains, both modes of fracture are effective, since the grain size ($d = 10$) is a bit larger than the indented depth. With the increase of grain size, the fracture mode is transformed from intergranular to transgranular fracture. Actually, the observations from Figure 3.20 confirm this transformation. For intergranular fracture, cracks propagate along grain boundaries and the fracture surface is always uneven. For transgranular fracture, however, cracks propagate within the grains and the fracture surface is smoother. This may be the reason for the difference in surface morphologies, as shown in Figure 3.20.

To model the plowing friction, σ_i and σ_t, as shown in Figure 3.26, are assumed to represent the intergranular and transgranular fracture strengths, respectively. The average plowing stress is designated as

$$\sigma_p = \lambda\sigma_t + (1-\lambda)\sigma_i \tag{3.54}$$

where λ is the proportion of the transgranular fracture. Apparently, λ is a function of grain size, d, and the indenting depth, h_p. For the cone-shaped rigid asperity, the maximum value of λ is

$$\lambda(d) = \begin{cases} \dfrac{d^2}{h_p^{\,2}} & d \le h_p \\[2mm] 1 & d > h_p \end{cases} \tag{3.55}$$

Obviously, for a given indenting depth, λ increases with grain size.

More generally, the grains with all sizes from small to large exist in a sample. This means both the fracture modes exist even in the samples with the largest grains. However, for those samples with a larger average grain size, the transgranular fracture takes a larger proportion, and λ has a higher value. The size of grains can be assumed to follow Gaussian distribution with the probability density function as

$$f(d) = \frac{1}{\sigma\sqrt{2\pi}} e^{-\frac{(d-d_a)^2}{2\sigma^2}} \tag{3.56}$$

where d_a is the average grain size and σ is the standard deviation of grain size. Its value is identified as $\sigma = \frac{d_a}{4}$. Thus, the proportion of the transgranular fracture for the whole surface, λ_s, is given by

$$\lambda_s = \int_{-\infty}^{+\infty} f(x)\lambda(x)dx = \int_{-\infty}^{h_p} \frac{4}{d_a\sqrt{2\pi}} e^{-\frac{8(x-d_a)^2}{d_a^2}} \frac{x^2}{h_p^2} dx + \int_{h_p}^{+\infty} \frac{4}{d_a\sqrt{2\pi}} e^{-\frac{8(x-d_a)^2}{d_a^2}} dx \tag{3.57}$$

According to the contact mechanics [47], the indenting depth depends on the mechanical properties of deformable material and the normal load. Considering that only half of the cone is in contact when sliding initiates, the indenting depth is determined as

$$\begin{cases} F^* = \eta \dfrac{p}{\sigma_0} \dfrac{\pi}{2}\left(\dfrac{h_p}{\cot\theta}\right)^2 \\ \dfrac{p}{\sigma_0} = \dfrac{2}{3}\left[1 - \ln\left(\dfrac{E\cot\theta}{3\sigma_0}\right)\right] \end{cases} \tag{3.58}$$

where p is the contact pressure on asperity. E and σ_0 are the Young's modulus and the yield stress of material. θ and η are the average semiangle and distribution density of rigid asperities.

The coefficient of plowing friction is thus obtained in the following:

$$\mu_p = \frac{F_p}{F_N} = \frac{3[\lambda_s\sigma_t + (1-\lambda_s)\sigma_i]\cot\theta}{\pi\sigma_0\left[1 - \ln\left(\dfrac{E\cot\theta}{3\sigma_0}\right)\right]} \tag{3.59}$$

when $\lambda_s = 1$, it is effective on grain boundary, let Eq. (3.59) equals to Eq. (3.50), then, the following equation is obtained:

$$\sigma_t = \frac{2}{3}\sigma_0\left[1 - \ln\left(\frac{E\cot\theta}{3\sigma_0}\right)\right] \tag{3.60}$$

If $\sigma_i = k\sigma_t$ ($k \in (0, 1)$) is a material constant, the coefficient of plowing friction can be represented as

$$\mu_p = \left[\lambda_s(1-k)+k\right]\frac{2}{\pi}\cot\theta \qquad (3.61)$$

According to Eq. (3.61), the coefficient of plowing friction increases with λ_s. Hence, the coefficient of plowing friction for the coarse-grained specimens is larger than that of the fine-grained specimens.

3.3.2.3 Comparison of the Models

As mentioned earlier, there are different kinds of plowing models proposed based on different mechanisms to explain the grain size effect on plowing friction coefficient. For example, the classical model is proposed based on the plastic deformation. Bucaille's model uses the elastic recovery to interpret the change of friction coefficient in the plowing process. Peng's model claims that the fracture failure is the main mechanism of plowing process. To determine the mechanism of the grain size effect on plowing process observed in Section 3.2.2.3, these three kinds of models with the experimental results are compared.

To actualize the comparison, the steel surface used in the experiments is simplified to be consisting of numerous conical asperities with the same height and semiangle. The semiangle of the rigid asperities is determined by measurement of the surface profile of the steel surface. The value of θ is calculated by the following:

$$\theta = \arcsin\left(\frac{\Delta x}{\int\sqrt{1+(y'(x))^2}\,dx}\right) \qquad (3.62)$$

where $y(x)$ is the profile of surface and Δx is the sampling length. In this experiment, the value of $\theta = 0.6$ is obtained. The value of $\eta = 200$ is also identified by counting the number of summits with the height greater than $16\,\mu m$, shown in Figure 3.20a.

By adopting the mechanical properties and the testing parameters listed in Table 3.2, the parameters λ_s and X of the different grain-sized specimens under different normalized normal loads are obtained and shown in Figure 3.27.

Figure 3.28 presents the plowing friction coefficient under the normalized normal load of 0.011 predicted by the present model (assuming $k = 0.8$) as well as the Bucallie's and Gorddard's models. It is obvious that the grain size effect on plowing friction, which is almost neglected by the previous models, is predicted by Peng's model successfully. Using Peng's model, the

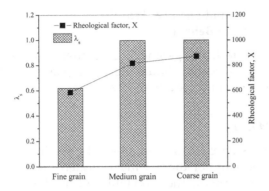

FIGURE 3.27
The value of λ_s and X as a function of grain size and normal load [4].

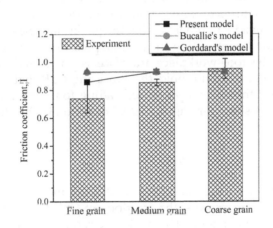

FIGURE 3.28
The predicted plowing friction coefficient under the normalized normal load of 0.011 [4].

predicted friction coefficient for the fine-grained samples is significantly smaller than those of other samples, and the previous models fail to describe this phenomenon.

According to Peng's model, the friction coefficient of the samples with medium-sized grains is the same as that of the coarse-grained sample, since λ equals to 1 for both samples in the testing condition. It is inconsistent with the experimental results, as an increase of friction coefficient is observed in the experiments from medium-sized grains to coarse-sized grains. This may be caused by the uneven distribution of grain size in the deformable sample. In fact, small grains exist in the samples even though the average grain size is more than 100 µm. Therefore, an intergranular fracture can happen even if the indenting depth is smaller than the average grain size. Since the average grain size increases, λ is close to but unable to reach 1 for polycrystalline alloys. The details need to be explored in future.

3.3.3 Geometry Size and Surface Asperity Affected Friction Behavior

Unlike grain size effect, the geometry size effect on friction behavior is investigated via the effect of the liquid lubricating cases. The lubrication thus needs to be considered in the study of this effect.

3.3.3.1 O/CLP Model

When a lubricant is applied to the tool–workpiece interface, some valleys among peaks, especially for those valleys located inside the interface, can hold the lubricant to form the enclosed lubricant, as illustrated in Figure 3.29. As a result, the friction force could be reduced because part of the normal pressure is offset by the enclosed lubricant. These valleys are named *CLPs*. On the other hand, some valleys, which are named *OLPs*, located on the outer part of the contact surface, cannot hold the lubricant because these valleys are open to the outside. In other words, the peaks and the closed lubricants make contribution to supporting the normal pressure on the contact surface.

Accordingly, the tool–workpiece interface is composed of three different parts: *OLP* which cannot hold the lubricant; *CLP* that can capture the lubricant to offset the normal pressure, leading to the decreases of friction force and the *RCA* shown in Figure 3.29. Equation (3.63) in the following is thus obtained:

$$\alpha_O + \alpha_C + \alpha_{RC} = 1 \qquad (3.63)$$

Here α_O, α_C, and α_{RC} are the fractions of *OLP*, *CLP*, and *RCA*, respectively.

During the forming process, the fractions of *CLP*, *OLP*, and *RCA* change with the displacement of tools. The peaks are compressed and α_{RC} increases, whereas α_O and α_C decrease. Figure 3.30 presents an image constructed based

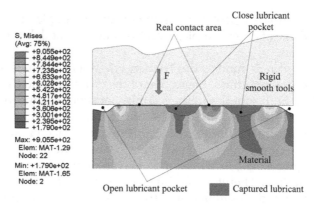

FIGURE 3.29
2D sketch of *OLPs*, *RCAs*, and *CLPs* [1].

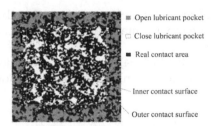

- ■ Open lubricant pocket
- □ Close lubricant pocket
- ■ Real contact area

- Inner contact surface
- Outer contact surface

FIGURE 3.30
Three fractions of *OLPs*, *CLPs*, and *RCAs* [1].

on the real topography data of metal sheet, which is tricolor coded to show the *OLP*, *CLP*, and *RCA* calculated areas for an imaginary plane [6]. According to the distribution of *OLP*, *CLP*, and *RCA*, the tool–workpiece interface is composed of two different areas: the inner contact area A_i and the outer contact area A_s, which are marked by dotted lines in the figure. It is obvious that there are more *CLPs* located in the inner contact surface than that in the outer contact surface, because the lubricant in the inside of valleys (in inner contact surface) is easily enclosed by adjacent peaks to generate *CLPs*.

Figure 3.31 presents a geometric model for inner and outer contact surface with minimization. Usually, the dimension parameter (s) of outer contact surface could be assumed to be equal, while D_0 and h_0 decreases with the size of the contact surface, as illustrated in Figure 3.31. In addition, the *OLPs* in the outer contact surface are much bigger than those in the inner contact surface, as shown in Figure 3.30. As a result, α_o will decrease with the size of the contact surface. Hence, with the decreasing size, the ratio of the outer contact surface increases until $A_s = A$, as shown in Figure 3.31. Consequently, the friction force in meso- and microscale could be much greater than that in the macroscale due to less normal pressure offset by *CLPs*. Therefore, macrofriction is not just a summation of various "zones" of microfriction despite of similar tribological features. To evaluate the ratio of the outer surface to the total contact surface, the scale factor is defined as follows:

$$\frac{A}{A_s} = \lambda \tag{3.64}$$

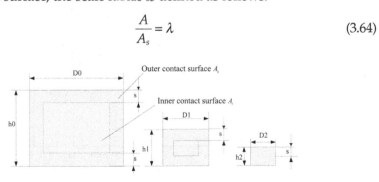

FIGURE 3.31
The change of outer and inner contact surface with the decrease of specimen size [1].

where A_s is the area of the outer contact surface, A is the area of the contact surface that includes the inner contact surface and outer contact surface, and λ is the size scale factor.

By combining Eqs. (3.63) and (3.64), the function relations of α_O, α_C, and α_{RC} considering the size scale factor can be obtained as follows [6]:

$$\frac{\alpha_c}{1 - \alpha_{RC}} = \frac{\alpha_c}{\alpha_c + \alpha_o} = 1 - \frac{1}{\lambda} \tag{3.65}$$

$$\frac{\alpha_o}{1 - \alpha_{RC}} = \frac{\alpha_o}{\alpha_c + \alpha_o} = \frac{2\lambda - 1}{\lambda^2} \tag{3.66}$$

Equation (3.65) is for double-side friction, while Eq. (3.66) is for single-side friction.

3.3.3.2 Friction Model in Meso- and Microscaled Plastic Deformation

The general friction model (WB model) is established on the basis of RCA. It is designated as follows [2]:

$$\tau = m\alpha_{RC}k \tag{3.67}$$

In Eq. (3.67), m is the friction factor (usually $0 \leq m \leq 1$); α_{RC} is the ratio of RCA in the tool–workpiece interface, as shown in Figure 3.29; k is the shear flow stress of material; and α_{RC} is the function of the normal pressure p/k with different friction factors m in dry condition, as shown in Figure 3.32.

The lubricant friction behavior is different from dry friction due to the influence of CLP. Figure 3.33a and c show the difference of the contact surface before and after the applied load in dry condition. In addition,

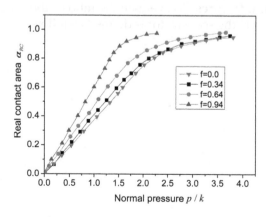

FIGURE 3.32

α_{RC} as the function of normal pressure p/k (dry friction) [1].

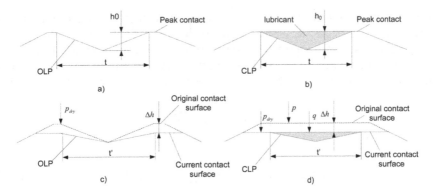

FIGURE 3.33
The deformation of materials in microscopic view [1].

Figure 3.33b and d show the situation in which the lubricant is used. By comparing the dry friction with the lubricant friction, it is found that the normal pressure (q) is reduced by the enclosed lubricant in *CLPs*. In meso- and microscale, the reduced force related to *CLPs*, can be calculated as $q\alpha_c A$. According to the force equilibrium condition in the normal direction of the contact surface, Eq. (3.68) in the following is thus established:

$$pA = p_{dry}A + q\alpha_C \cdot A \qquad (3.68)$$

where p_{dry} is the normal pressure for dry friction relating to *RCA* in Wanheim/Bay general friction model and q is the pressure reduced by the enclosed lubricant in *CLPs*.

Therefore, Eq. (3.68) can further be expressed as

$$p = p_{dry} + (1 - \alpha_{RC} - \alpha_o)q \qquad (3.69)$$

In addition, each tiny CLP can be supposed as some kind of triangle shape, as shown in Figure 3.33b. In this study, it is assumed that the tool is a rigid body and its surface is smooth because the tool material is much harder than the specimen material. For tool displacement Δh and the trapped lubricant pressure q, Eq. (3.70) could be obtained [48].

$$\begin{cases} \dfrac{q}{K_0} = \dfrac{1}{K_1}\left[\left(1 - \dfrac{2\Delta h}{h_0}\right)^{-K_1} - 1\right] \\[3mm] \dfrac{\Delta h}{h_0} = \dfrac{3+C}{2(2+C)}\alpha_{RC} \\[3mm] C = \dfrac{3\tan\gamma_0 - 4 + 2(3\tan\gamma_0 + 4)^{1/2}}{4 - \tan\gamma_0} \end{cases} \qquad (3.70)$$

where K_0 and K_1 are the antipressure strength parameters related to lubricant on the contact surface; K_1 is the higher-order coefficient of the unload function; γ_0 is the asperity slope angle related to the profile of the material surface.

By substitution Eqs. (3.65) and (3.66) into Eq. (3.69), the function equations could be derived between the normal pressure, size scale factor, and *RCA*.

$$p = p_{dry} + \left[(\lambda - 1) \cdot \frac{1 - \alpha_{RC}}{\lambda}\right] q \tag{3.71}$$

$$p = p_{dry} + \left[1 - \alpha_{RC} - \frac{2\lambda - 1}{\lambda^2}(1 - \alpha_{RC})\right] q \tag{3.72}$$

In Eqs. (3.71) and (3.72), p_{dry} is the normal pressure for dry friction, which could be obtained by Wanheim/Bay general friction model. To consider the influence of *CLPs* on *RCA* in lubricant friction, size scale factor λ is introduced. Equation (3.71) is for double-side friction and Eq. (3.72) is for single-side friction. Equations (3.71) and (3.72) describe the evolvement of the force balance from dry friction ($\lambda = 1$) to macroscale lubricant friction ($\lambda \to \infty$).

To obtain a dimensionless normal pressure, Eqs. (3.71) and (3.72) are divided by the yield stress of the material σ_0, respectively, and Eqs. (3.73) and (3.74) are obtained:

$$\frac{p}{\sigma_0} = \frac{p_{dry}}{\sigma_0} + \left[(\lambda - 1) \cdot \frac{1 - \alpha_{RC}}{\lambda}\right] \frac{q}{\sigma_0} \tag{3.73}$$

$$\frac{p}{\sigma_0} = \frac{p_{dry}}{\sigma_0} + \left[1 - \alpha_{RC} - \frac{2\lambda - 1}{\lambda^2}(1 - \alpha_{RC})\right] \frac{q}{\sigma_0} \tag{3.74}$$

Therefore, *RCA* α_{RC} can be expressed by the following equations:

$$\alpha_{RC} = \varphi\left(\frac{p_{dry}}{\sigma_0}, m\right) = \varphi\left(\frac{p - \left[(1 - \lambda) \cdot \frac{1 - \alpha_{RC}}{\alpha_{RC}\lambda}\right] q}{\sigma_0}, m\right) \tag{3.75}$$

$$\alpha_{RC} = \varphi\left(\frac{p_{dry}}{\sigma_0}, m\right) = \varphi\left(\frac{p - \left[\frac{1 - \alpha_{RC}}{\alpha_{RC}} - \frac{2\lambda - 1}{\lambda^2 \alpha_{RC}}(1 - \alpha_{RC})\right] q}{\sigma_0}, m\right) \tag{3.76}$$

where φ depicts the α_{RC} as the function of normal pressure p and size scale factor λ. By using this equation, the RCA α_{RC} with the consideration of size effect could be figured out. According to Eq. (3.66), the friction force from macro- to microscale could be calculated by Eq. (3.77), and the new uniform fiction model is established for multiscale.

$$\tau = m\varphi(p, \lambda)k \qquad (\lambda \geq 1) \tag{3.77}$$

When size scale factor changes from 1 to $+\infty$, the friction decreases from dry friction to macrofriction. Therefore, the friction in meso- and microscale is between the dry friction-the upper boundary and the conventional macro lubricant friction-lower boundary. It can be expressed by

$$\tau_{dry} \geq \tau_{mic} \geq \tau_{mac} \tag{3.78}$$

where τ_{dry} is the friction force of dry friction, τ_{mic} is the friction force of meso- and microscale friction considering size effect, τ_{mac} is the friction force of macrolubricant friction. They are expressed by Eq. (3.79).

$$\begin{cases} \tau_{dry} = m\varphi(p, 1)k \\ \tau_{mic} = m\varphi(p, \lambda)k \\ \tau_{mac} = m\varphi(p, \infty)k \end{cases} \tag{3.79}$$

3.3.3.3 FE Simulation of Friction Behaviors in Meso- and Microforming

The earlier friction model is integrated with the FE simulation model to study the friction size effect in meso- and microscale ring compression process. The geometric dimensions are listed in Table 3.3.

In addition, Figure 3.34 shows the schematic of the simulation model for ring compression test. It consists of two rigid tools (the upper movable tool and the stationary tool) with a ring material between them. Figure 3.35 shows the strain–stress curves of the material in different scales obtained from Messner's experiments [10].

TABLE 3.3

Parameters of the Ring Compression Process Analysis [1]

No	Initial Height l_0 (mm)	Initial Outer Diameter m_0 (mm)	Initial Inner Diameter n_0 (mm)	Geometric Dimension Scale Factor β	Friction Scale Factor λ
1	1.68	4.8	2.4	1	6
2	0.70	2.0	1.0	0.42	2.5
3	0.35	1.0	0.5	0.21	1.25

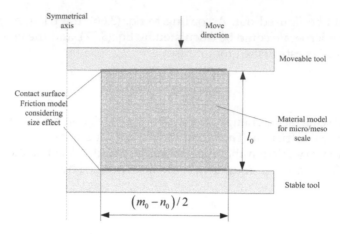

FIGURE 3.34
Schematic of numerical model for ring compression test [1].

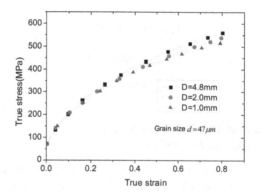

FIGURE 3.35
True strain–stress curves in different scales [1].

Figure 3.36 illustrates the friction behaviors predicted by the proposed model (RAC as the function of normal pressure and different scale factor). To explain the size effect on friction observed in ring compression process, we need to implement the model into the commercial FE analysis software (ABAQUS-Standard). Some modifications should be done to make it feasible for numerical simulation. From Figure 3.36, the friction model with size effect can be taken as $\dfrac{\tau}{m \cdot k} \leftrightarrow \left(\dfrac{p}{\sigma_0}, \lambda\right)$. According to Eq. (3.77), the friction force of various scales can be obtained as follows.

$$\tau = \left(\frac{m \cdot k}{p} \varphi\left(\frac{p}{\sigma_0}, \lambda\right)\right) \cdot p \tag{3.80}$$

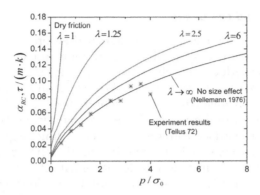

FIGURE 3.36
Friction behaviors in different scales ($K_0/\sigma_0 = 11.3$, $K_1 = 9.25$, $\gamma = 15°$, $m = 0.94$) [1].

In Eq. (3.80), φ is α_{RC} as a function of normal pressure for a specific λ, which can be calculated from the friction curves in Figure 3.36. Using this transformation, the new friction model is transformed to a special Coulomb friction model, and the first term of the equation $\left(\dfrac{m \cdot k}{p} \Psi\left(\dfrac{p}{\sigma_0}, \lambda\right)\right)$ can be taken as the friction coefficient μ for the traditional Coulomb friction.

According to the method mentioned earlier, friction coefficient at different normal pressures for the fixed size scale factors can be calculated as listed in Table 3.4. As a result, the friction model can be easily input into the FE models established in ABAQUS-Standard according to the traditional Coulomb friction.

TABLE 3.4

Friction Coefficient for Different Normal Pressures with Various Size Scale Factors [1]

Normal Pressure (MPa)	$\lambda = 1.25$	$\lambda = 2.5$	$\lambda = 6$	$\lambda \rightarrow \infty$ No Size Effect
50	0.213	0.111	0.086	0.071
100	0.144	0.083	0.067	0.055
150	0.113	0.069	0.056	0.047
200	0.095	0.061	0.049	0.041
250	0.082	0.054	0.045	0.037
300	0.073	0.049	0.041	0.034
350	0.066	0.045	0.037	0.031
400	0.061	0.041	0.035	0.029
450	0.055	0.039	0.032	0.027
500	0.052	0.036	0.031	0.025

3.3.3.4 Analysis and Discussion

The plane stress FE models can be established according to different geometric dimensions, material model, and friction model for different size scales. The height reduction such as 60% is generally used in experiments. The radial friction force and the change of the inner and outer ring diameters are calculated to investigate the influence of friction size effect.

In addition, a series of simulations have been conducted. The radial friction force on the contact surface with the reduction of ring height is presented in Figure 3.37. For $D_0 = 1.0$, $\lambda = 1.25$, the maximum friction force is 86.1N without size effect. However, it is increased to 162.181N with the consideration of size effect, which is almost twice the value without considering the size effect. For $D_0 = 1.0$, $\lambda = 6$, the maximum friction force is 540.9N without considering size effect and 650.5N with size effect being considered, respectively. The difference becomes smaller than that for $D_0 = 1.0$, $\lambda = 1.25$ despite the increased variation magnitude. It means that the size effect in friction will be more obvious with the reduction of size.

In addition to the friction force, the changes of inner and outer compressed ring diameters are usually related to friction behaviors. Figures 3.38 and 3.39 illustrate the inner and outer diameter change with the ring height reduction. It is found that the inner diameters increase in three different scales, without consideration of size effects, as shown in Figure 3.38. Once the size effect is taken into consideration, different results are obtained. For $D_0 = 1.0$, $\lambda = 1.25$, the inner diameter is decreased by 21.92% when the height reduction reaches 60%; for $D_0 = 2.0$, $\lambda = 2.5$, the inner diameter is increased at the beginning and then decreased by 5.4% at the end; for $D_0 = 4.8$, $\lambda = 6$, the inner diameter is increased by 13.93%. The reason is that the friction in small scale is much greater than that in large scale. The difference of the increasing

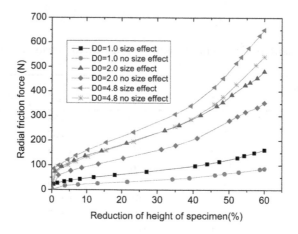

FIGURE 3.37
Radial friction force versus ring height reduction [1].

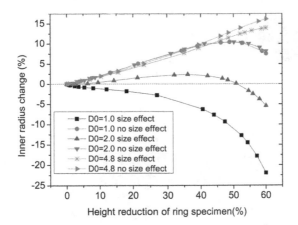

FIGURE 3.38
Inner diameter change versus ring height reduction [1].

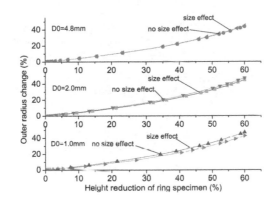

FIGURE 3.39
Outer diameter change versus ring height reduction [1].

outer diameters, with and without size effect being considered, is greater in small scale ($D_0 = 1.0$, $\lambda = 1.25$). The smaller scale, the greater friction, is caused by the greater influence of size effect.

According to the scaling law, the Mises stress distribution of ring part after compression and radial displacement of the compressed ring for various scales are obtained and presented in Figure 3.40 by geometric transformation, viz., multiplying $1/\lambda$. The red dot line in the figure is the initial position of inner radius. By comparison of the compressed ring parts with the height reduction of 60% in various size scales, the influence of friction size effect in ring compression process can be obviously observed. Furthermore, the results are quite consistent with Messner's experiments [49].

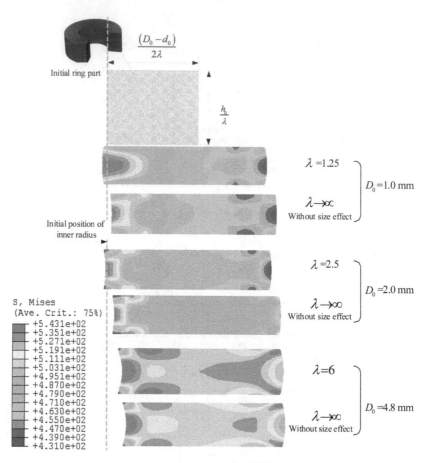

FIGURE 3.40
Ring parts after compression and radial displacements in various scales [1].

3.4 Summary

The chapter presents the experimental observations of geometry and grain size effects on the friction behavior in meso- and microforming processes. The geometry size effect only affects friction under liquid lubricating conditions, whereas the tooling–workpiece interfacial friction increases with the decrease of specimen size. It is caused by the increasing fraction of OLP, which is unable to retain liquid lubrication. The asperities around OLPs need to maintain a greater normal pressure and more RCA is thus generated by asperity flattening. The observation of grain size effect on friction needs to handle the adhesion and plowing mechanism separately. Friction processes are generally dominated by different mechanisms, but the grain size has an opposite effect. It is observed experimentally that the increase of grain size results in the decrease of coefficient of adhesive friction. But the coefficient of plowing friction increases with grain size. With the increase of normal load, the mechanism of friction is transformed from adhesion to plowing. The main mechanism of grain size effect on adhesive friction is the increasing proportion of plastic deformation with the increase of grain size. The mechanical property that affects the friction coefficient is the ratio of $\dfrac{E}{H}$ or $\dfrac{E}{\sigma_0}$, while the main mechanism of plowing friction is considered to be the transformation of transgranular and intergranular fractures of the softer material with the increase of grain size.

References

1. L. Peng, X. Lai, H.-J. Lee, J.-H. Song, J. Ni, Friction behavior modeling and analysis in/meso scale metal forming process, *Materials & Design* 31 (2010) 1953–1961.
2. T. Wanheim, N. Bay, A.S. Petersen, A theoretically determined model for friction in metal working processes, *Wear* 28 (1974) 251–258.
3. F. Vollertsen, H. Schulze Niehoff, Z. Hu, State of the art in micro forming, *International Journal of Machine Tools and Manufacture* 46 (2006) 1172–1179.
4. L.F. Peng, M.Y. Mao, M.W. Fu, X.M. Lai, Effect of grain size on the adhesive and ploughing friction behaviours of polycrystalline metals in forming process, *International Journal of Mechanical Sciences* 117 (2016) 197–209.
5. F. Vollertsen, Z. Hu, H.S. Niehoff, C. Theiler, State of the art in micro forming and investigations into micro deep drawing, *Journal of Materials Processing Technology* 151 (2004) 70–79.
6. U. Engel, Tribology in microforming, *Wear* 260 (2006) 265–273.
7. F. Vollertsen, Z. Hu, Tribological size effects in sheet metal forming measured by a strip drawing test, *CIRP Annals – Manufacturing Technology* 55 (2006) 291–294.
8. M.F. Ashby, The deformation of plastically non-homogeneous materials, *Philosophical Magazine* 21 (1970) 399–424.
9. W.L. Chan, M.W. Fu, J. Lu, The size effect on micro deformation behaviour in micro-scale plastic deformation, *Materials & Design* 32 (2011) 198–206.
10. A. Buschhausen, K. Weinmann, J.Y. Lee, T. Altan, Evaluation of lubrication and friction in cold forging using a double backward-extrusion process, *Journal of Materials Processing Technology* 33 (1992) 95–108.
11. N. Tiesler, U. Engel, M. Geiger, Forming of microparts—Effects of miniaturization on friction, in *6th International Conference on Technology of Plasticity (ICTP)* 1999 Nuremberg, Germany Berlin, Springer, 19–24 Sep. 1999, Vol. 11, pp. 889–894.
12. M. Geiger, M. Kleiner, R. Eckstein, N. Tiesler, U. Engel, Microforming, *CIRP Annals – Manufacturing Technology* 50 (2001) 445–462.
13. G. Bin, G. Feng, C.-J. Wang, D.-B. Shan, Flow stress and tribology size effects in scaled down cylinder compression, *Transactions of Nonferrous Metals Society of China* 19 (2009) s516–s520.

14. R. Ebrahimi, A. Najafizadeh, A new method for evaluation of friction in bulk metal forming, *Journal of Materials Processing Technology* 152 (2004) 136–143.
15. J.H. Deng, M.W. Fu, W.L. Chan, Size effect on material surface deformation behavior in micro-forming process, *Materials Science and Engineering: A* 528 (2011) 4799–4806.
16. L.F. Mori, N. Krishnan, J. Cao, H.D. Espinosa, Study of the size effects and friction conditions in microextrusion—Part II: Size effect in dynamic friction for brass-steel pairs, *Journal of Manufacturing Science and Engineering* 129 (2007) 677–689.
17. A. Moshkovich, V. Perfilyev, I. Lapsker, D. Gorni, L. Rapoport, The effect of grain size on Stribeck curve and microstructure of copper under friction in the steady friction state, *Tribology Letters* 42 (2011) 89–98.
18. A. Moshkovich, V. Perfilyev, I. Lapsker, L. Rapoport, Stribeck curve under friction of copper samples in the steady friction state, *Tribology Letters* 37 (2010) 645–653.
19. T. Senda, E. Yasuda, M. Kaji, R.C. Bradt, Effect of grain size on the sliding wear and friction of alumina at elevated temperatures, *Journal of the American Ceramic Society* 82 (1999) 1505–1511.
20. D. Tabor, Junction growth in metallic friction: The role of combined stresses and surface contamination, *Proceedings of the Royal Society of London. Series A. Mathematical and Physical Sciences* 251 (1959) 378.
21. J. Greenwood, J. Tripp, The contact of two nominally flat rough surfaces, *Proceedings of the Institution of Mechanical Engineers* 185 (1970) 625–633.
22. M. Mao, L. Peng, P. Yi, X. Lai, Modeling of the friction behavior in metal forming process considering material hardening and junction growth, *Journal of Tribology* 138 (2016) 012202.
23. J. Greenwood, J. Williamson, Contact of nominally flat surfaces, *Proceedings of the Royal Society of London. Series A. Mathematical and Physical Sciences* 295 (1966) 300–319.
24. Y. Zhao, D.M. Maietta, L. Chang, An asperity microcontact model incorporating the transition from elastic deformation to fully plastic flow, *Journal of Tribology* 122 (2000) 86–93.
25. W. Chang, I. Etsion, D.B. Bogy, An elastic-plastic model for the contact of rough surfaces, *Journal of Tribology* 109 (1987) 257–263.
26. P. Follansbee, G. Sinclair, Quasi-static normal indentation of an elasto-plastic half-space by a rigid sphere—I: Analysis, *International Journal of Solids and Structures* 20 (1984) 81–91.
27. G. Sinclair, P. Follansbee, K. Johnson, Quasi-static normal indentation of an elasto-plastic half-space by a rigid sphere—II. Results, *International Journal of Solids and Structures* 21 (1985) 865–888.
28. R.L. Jackson, I. Green, A finite element study of elasto-plastic hemispherical contact against a rigid flat, *Journal of Tribology* 127 (2005) 343–354.
29. L. Kogut, I. Etsion, Elastic-plastic contact analysis of a sphere and a rigid flat, *Journal of Applied Mechanics* 69 (2002) 657–662.
30. W. Johnson, Extrusion through square dies of large reduction, *Journal of the Mechanics and Physics of Solids* 4 (1956) 191–198.
31. T. Wanheim, N. Bay, A. Petersen, A theoretically determined model for friction in metal working processes, *Wear* 28 (1974) 251–258.

32. D. Tabor, Junction growth in metallic friction: the role of combined stresses and surface contamination, *Proceedings of the Royal Society of London. Series A. Mathematical and Physical Sciences* 251 (1959) 378–393.
33. M.M. Chaudhri, The junction growth equation and its application to explosive crystals, *Journal of Materials Science Letters* 3 (1984) 565–568.
34. F.P. Bowden, D. Tabor, *The Friction and Lubrication of Solids*, vol. 1, Oxford University Press, Great Clarendon Street, Oxford, 2001.
35. A. Majumdar, B. Bhushan, Characterization and modeling of surface roughness and contact mechanics, in *Handbook of Nano-Micro-Tribology*, Editor B. Bushan, CRC Press, Boca Raton, FL, 1995.
36. L. Kogut, I. Etsion, A finite element based elastic-plastic model for the contact of rough surfaces, *Tribology Transactions* 46 (2003) 383–390.
37. L. Kogut, I. Etsion, A static friction model for elastic-plastic contacting rough surfaces, *Journal of Tribology* 126 (2004) 34–40.
38. W.-R. Chang, I. Etsion, D. Bogy, Adhesion model for metallic rough surfaces, *Journal of Tribology* 110 (1988) 50–56.
39. D. Cohen, Y. Kligerman, I. Etsion, A model for contact and static friction of nominally flat rough surfaces under full stick contact condition, *Journal of Tribology* 130 (2008) 031401.
40. L. Li, I. Etsion, F. Talke, Contact area and static friction of rough surfaces with high plasticity index, *Journal of Tribology* 132 (2010) 031401.
41. J. Goddard, H. Wilman, A theory of friction and wear during the abrasion of metals, *Wear* 5 (1962) 114–135.
42. J. Bressan, G. Genint, J. Williams, The influence of pressure, boundary film shear strength and elasticity on the friction between a hard asperity and a deforming softer surface, *Tribology Series* 36 (1999) 79–90.
43. J. Bucaille, E. Felder, G. Hochstetter, Mechanical analysis of the scratch test on elastic and perfectly plastic materials with the three-dimensional finite element modeling, *Wear* 249 (2001) 422–432.
44. P.L. Gutshall, G.E. Gross, Observations and mechanisms of fracture in polycrystalline alumina, *Engineering Fracture Mechanics* 1 (1969) 463–471.
45. Z. Xu, L. Peng, M. Fu, X. Lai, Size effect affected formability of sheet metals in micro/meso scale plastic deformation: Experiment and modeling, *International Journal of Plasticity* 68 (2015) 34–54.
46. B. Meng, M. Fu, C. Fu, J. Wang, Multivariable analysis of micro shearing process customized for progressive forming of micro-parts, *International Journal of Mechanical Sciences* 93 (2015) 191–203.
47. K.L. Johnson, *Contact Mechanics*, Cambridge University Press, Cambridge, 1987.
48. T. Nellemann, N. Bay, T. Wanheim, Real area of contact and friction stress—The role of trapped lubricant, *Wear* 43 (1977) 45–53.
49. A. Messner, U. Engel, R. Kals, F. Vollertsen, Size effect in the FE-simulation of micro-forming processes, *Journal of Materials Processing Technology* 45 (1994) 371–376.

4

Instability and Formability of Materials in Meso- and Microforming

4.1 Introduction

As discussed and presented in the previous chapters, the size effect has been proven to affect the performance of meso- and microforming processes, evidently indicated by the variation of the flow stress of working materials, friction in the forming system, deformation load, etc. As a consequence, the traditionally established theories may not be fully valid and efficient in these downscaled forming processes, let alone to support the micropart design, microforming process determination, process route identification, and process parameter configuration, and further for product quality control and assurance. These eluded issues remain unrevealed. In the past two decades, more efforts from academia and industries have been provided to address these issues and to explore an in-depth insight into the size effects and their affected process performance and product quality in microscaled manufacturing [1].

Nevertheless, the size effect on the instability and failure behavior of materials in meso- and microforming is another critical issue to be addressed and has been becoming more crucial due to the overwhelming trend of product miniaturization and the promising application potential of meso- and microforming processes. Therefore, many pioneering research projects were conducted. Among them, the pneumatic bulge tests of aluminum alloys were done by Vollertsen et al. [2], and the highly irregular local distribution of strain was found to exist in microdeformation. The fracture was also shown to occur randomly within the deformation zone. The uniaxial tensile tests of the annealed copper foils with different thicknesses and grain sizes were carried out by Fu and Chan [3]. They found that the fracture stress and strain as well as the number of microvoids decrease with the thickness-to-grain size ratio (t/d). A dislocation density based model considering the size effect was thus developed to analyze the deformation, and the analytical results were verified by experiments. Furthermore, Ben Hmida et al. [4] investigated the single-point increment-forming process of copper foils with different grain sizes. It is shown that the formability of specimens deteriorates with the reduction of t/d. Furushima et al. [5] studied the fracture and free surface roughening behavior of copper foils during the uniaxial tensile test. They reported that the fracture strain decreases with the thickness of the foil specimen. A significant increase in surface roughness was also observed with the decrease of foil thickness. Moreover, a thorough investigation of the ductile fracture (DF) of micro- and macroscale flanged upsetting of brass specimen was performed by Ran et al. [6]. The fracture on the flanged surface is shown to form more easily in macroscale compared with the microscale deformation. They also pointed out that step-like elongated shear dimples exist in the two fracture surfaces of the broken flanged part. The dimple size on the transgranular fracture surfaces is affected by different scaling factors. The occurrence of these fracture behaviors is argued to be caused greatly by size effect. By describing the effects of dislocation density and phase volume fraction on the flow stress of materials in microscaled deformation process, a hybrid model considering multiphase of material and size effect was further proposed by them to predict the fracture in microforming [7]. The in-depth relationship of grain size effect and geometry size effect in microscale DF deformation was also investigated and analyzed. In addition, the applicability of different DF criteria was also discussed based on the established flow stress model in their further study [8]. The Freudenthal criterion was revealed to be suitable for the analysis of DF in compression-dominant deformation processes in both macro- and microscales. Furthermore, Meng et al. [9,10] investigated the DF and deformation behavior in the progressive microforming and shearing processes. They found that the fracture surface, inhomogeneous deformation, and irregular geometry deteriorate with the increase of grain size, which is caused by the accumulative effect of each forming operation. Different traditional fracture criteria were also employed to verify their applicability in the microforming process. They pointed out that each criterion has its own limitation in terms of the applicable range of workpiece dimension and stress state. Therefore, none of the fracture criteria can predict the fracture behavior well for both the microshearing and blanking operations in the progressive microforming process.

From the earlier review, it can be seen that most research on the instability and formability of materials in meso- and microforming processes are focused on one or several particular microforming processes. There is still a lack of in-depth and systematic study and understanding on the behavior and mechanism of meso- and microscaled necking and fracture under different loading conditions. In this chapter, the size effect on the failure behavior of sheet metals under different loading paths is analyzed and presented based on a thorough experimental investigation on the meso- and microforming limit of sheet metals with different thicknesses and grain sizes.

4.2 Forming Limit Measurement by Experiments

Regarding the failure of ductile materials, there are many unsolved problems and unrevealed mechanisms for the necking and DF behaviors. So far, there is still no unique and proven theory that can explain all the phenomena of ductile failure as well, as how and why they happen and behave. Therefore, this issue is still attracting the attention of researchers from academia and practitioners from industries all over the world. The forming limit diagram (FLD) proposed by Keeler [11] and Goodwin [12] has been widely employed in industries to evaluate and characterize the formability of sheet metals. FLD can not only describe the necking limit strain under simple constant load paths, it is also one of the most successful and widely accepted methods to articulate the forming limit of sheet metals at macroscale due to its concise and intuitive form as well as simple and convenient application. When the sheet metals are scaled down to meso- and microscale, how to employ the traditional FLD method to reveal the formability of sheet metals at meso- and microforming on the basis of an in-depth understanding of the fundamental mechanisms of size effect on the meso- and microscaled failure behaviors is thus an attractive research topic.

4.2.1 Material Characterization

To characterize the forming limit of sheet metals under different geometry and grain size conditions, the pure copper (Cu-FRHC) specimens with different thicknesses of 0.1, 0.2, and 0.4 mm were first prepared. The specimens were heat treated under different conditions to obtain different grain size conditions. The heat treatment parameters are summarized in Table 4.1. The microstructures of the specimens are shown in Figure 4.1.

TABLE 4.1

Heat Treatment Parameters [13]

Thickness (mm)	Temperature (°C)	Dwelling Time (h)	Average Grain Size (μm)	Grain Size Deviation (μm)
0.4	640	0.5	23.7	1.8
	720	0.5	58.9	12.6
	790	0.5	132.2	19.2
0.2	530	2	17.4	0.5
	620	2	35.2	3.7
	690	2	166.3	23.7
0.1	620	0.5	27.5	1.0
	720	0.5	48.3	8.7
	850	0.5	90.7	17.4

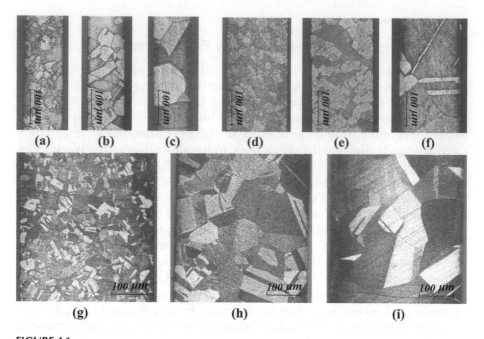

FIGURE 4.1
Microstructures of the specimens after different heat treatments [13]: (a) $t = 0.1$, $d = 27.5$; (b) $t = 0.1$, $d = 48.3$; (c) $t = 0.1$, $d = 90.7$; (d) $t = 0.2$, $d = 17.4$; (e) $t = 0.2$, $d = 35.2$; (f) $t = 0.2$, $d = 166.3$; (g) $t = 0.4$, $d = 23.7$; (h) $t = 0.4$, $d = 58.9$; and (i) $t = 0.4$, $d = 132.2$. (The units of t and d are mm and μm, respectively).

The uniaxial tensile tests were first employed to characterize the deformation and failure behavior of specimens under different geometry and grain size conditions. The engineering strain versus engineering stress curves were first obtained and shown in Figure 4.2.

From Figure 4.2, it is found that there is a significant decrease of flow stress and fracture strain with the increase of grain size. The reduction of fracture strain demonstrates the increasing deterioration trend of formability with the increase of grain size under uniaxial tensile condition. On the other hand, the fracture strain obtained is not the localized strain at the fracture spot, but an averaged strain over the gauge length. Therefore, the major reason for the decrease of fracture strain could be attributed to the inhomogeneous deformation of plastic strain along the gauge length, with the grain size getting close to the thickness of specimen. As a result, the strain is easier to localize at the weak region leading to an early fracture.

The decrease of flow stress with the increase of t/d has been well explored on the basis of surface layer model [14,15]. As explained by that model, the deforming polycrystalline sheet metal is divided into surface and internal grains. Since the free surface facilitates the rotation and deformation of grains, the flow stress on the surface grains becomes smaller than that

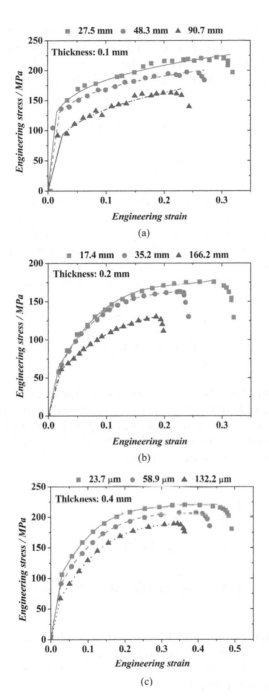

FIGURE 4.2
Engineering strain versus engineering stress curves of the specimens with different thicknesses and grain sizes: (a) $t = 0.1$ mm; (b) $t = 0.2$ mm; and (c) $t = 0.4$ mm.

of the internal ones. As the grain size increases, the surface grains take a greater proportion, which leads to a reduction of overall flow stress. A composited constitutive model was developed in [15] to describe the size effect of flow stress. According to the model, the flow stress can be formulated as

$$
\begin{cases}
\bar{\sigma} = \eta\bar{\sigma}_s + (1-\eta)\bar{\sigma}_i \\
\eta = \dfrac{N_s}{N}
\end{cases}
\tag{4.1}
$$

In Eq. (4.1), $\bar{\sigma}$ and N are the flow stress and grain number of the specimen, $\bar{\sigma}_s$ and N_s are the flow stress and number of surface grains, and $\bar{\sigma}_i$ is the flow stress of internal grains. The surface grains are considered to be similar to a single crystal, while the internal grains are taken as polycrystal. Crystal plastic theory and Hall–Petch equation [16,17] are employed to describe the stresses of surface and internal grains, respectively, as follows:

$$
\begin{cases}
\bar{\sigma}_s(\varepsilon) = m\tau_R(\varepsilon) \\
\bar{\sigma}_i(\varepsilon) = M\tau_R(\varepsilon) + \dfrac{k(\varepsilon)}{\sqrt{d}}
\end{cases}
\tag{4.2}
$$

In Eq. (4.2), m and M are the orientation factors of single crystal and polycrystal, respectively. τ_R is the critical-resolved shear stress of single crystal, k is the locally intensified stress needed to propagate general yield stress across the polycrystal grain boundaries, and d is the average grain size. Considering Eq. (4.1), the flow stress can be designated as

$$
\begin{cases}
\bar{\sigma} = \sigma_{\text{ind}} + \sigma_{\text{dep}} \\
\sigma_{\text{ind}} = M\tau_R(\varepsilon) + \dfrac{k(\varepsilon)}{\sqrt{d}} \\
\sigma_{\text{dep}} = \eta\left(m\tau_R(\varepsilon) - M\tau_R(\varepsilon) - \dfrac{k(\varepsilon)}{\sqrt{d}} \right)
\end{cases}
\tag{4.3}
$$

In Eq. (4.3), σ_{ind} and σ_{dep} represent the size-independent and -dependent parts of the flow stress, respectively. For sheet metals, the size factor η can be determined as the ratio between the surface grain number of a sheet specimen and the total grain number denoted as

$$
\eta = \frac{N_s}{N} = \frac{(w+t)/d - 2}{wt/d^2} = \frac{d}{t}\frac{w+t-2d}{w}
\tag{4.4}
$$

TABLE 4.2

The Fitted Results of Peng's Hybrid Model [13]

Parameters	0.1 mm	0.2 mm	0.4 mm
k_1	162.58	142.19	168.01
n_1	0.75	0.64	0.62
B	47.60	17.24	17.35
k_2	20.00	12.90	24.00
n_2	0.10	0.24	0.30

In meso- and microforming, the specimen width w is usually much larger than the sheet thickness t and the grain size d; hence, Eq. (4.4) can be simplified as

$$\eta = \frac{d}{t} \tag{4.5}$$

In addition, $k(\bar{\varepsilon}_m)$ and $\tau_R(\bar{\varepsilon}_m)$ are assumed to follow the following equations:

$$\begin{cases} k(\bar{\varepsilon}_m) = k_1(\bar{\varepsilon}_m)^{n_1} + b \\ \tau_R(\bar{\varepsilon}_m) = k_2(\bar{\varepsilon}_m)^{n_2} \end{cases} \tag{4.6}$$

By using the least-squares method, the parameters can be determined by fitting to the experimental true stress–strain curves. The results are shown in Figure 4.3 and Table 4.2.

Peng's model can well characterize the size effect affecting the constitutive behavior of plastic deformation under different thickness and grain conditions according to the calculation results. Therefore, the model was further employed in Section 4.4.4, where the analysis and modeling of failure behaviors at meso- and microscales are presented.

4.2.2 Forming Limit Experiments

There are plenty of methods to obtain the FLD of sheet metals at macroscale, such as the Nakazima test [18], biaxial tensile test [19], hydroforming test [20], etc. Nevertheless, not all the methods are fully suitable for the investigation of meso- and microscale forming limit. There are several issues that must be considered beforehand. At first, the effect of friction should be avoided in the experiments, considering that the friction behavior has been proven to be influenced by the size effect at meso- and microscale [21]. In addition, the proper miniaturization of apparatus dimensions needs to be conducted to simulate the deformation at meso- and microscale. The scale factor, on the other hand, needs to be well designed considering the feasibility of fabrication and assembly based on error analysis. Furthermore, a new strain measurement method also needs to be developed. The traditional method

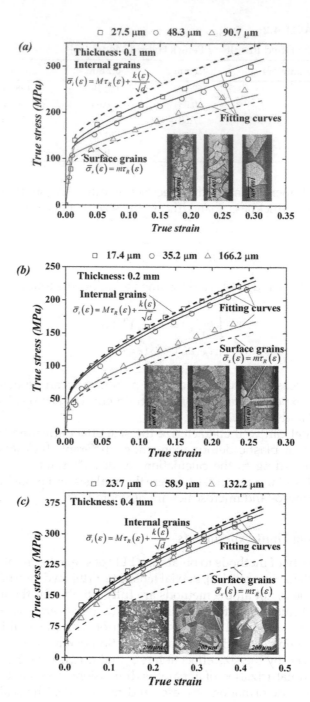

FIGURE 4.3
True stress–strain curves of the specimens with different thicknesses fitted based on Peng's constitutive model [13]: (a) $t = 0.1\,mm$; (b) $t = 0.2\,mm$; and (c) $t = 0.4\,mm$.

is to measure the dimensions of preprinted patterns on the specimen before and after deformation. The method, however, is difficult to realize as the size is scaled down to meso- and microlevel, and much more miniaturized and accurate patterns need to be developed and measured.

4.2.2.1 Experiment Design

Considering the previous issues, the Holmberg [22] and Marciniak (ISO 12004-2: 2008) methods were employed to investigate the forming limit of sheet metals under different loading conditions as shown in Figures 4.4 and 4.5. By stretching the proper-designed specimens of different widths with uniaxial loading, the left side of the forming limit curve (FLC) can be developed based on the Holmberg test. The Holmberg test excludes the effect of friction. The preparation of apparatus and specimen is also easy to conduct for different scales with a satisfying accuracy due to the simple formation. In addition, the Marciniak method was employed to obtain the right side of FLC by using a driving plate with a hole in the middle. During the test, both the plate and the specimen are subjected to deep drawing with a cylinder punch. The proper-designed specimen is hereby stretched by the tightly compressed driving plate until the occurrence of fracture in the middle area. This method also avoids the influence of friction and is relatively easy to realize at mesoscale.

To obtain the reliable limit strain results under different dimensional scales, the digital image correlation (DIC) method was employed for measurement of the limit strains in Holmberg and Marciniak tests, as illustrated in Figure 4.5. The specimens were first painted with random black and white patterns. During the deformation process, continuous digital images were obtained every 2 s until the fracture of the workpiece. By following the evolution of the recorded random patterns, the displacement and strain field can be calculated.

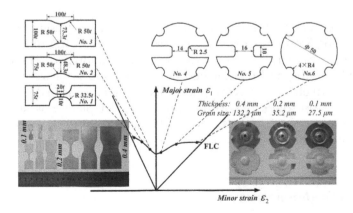

FIGURE 4.4
Design of the specimen geometry [13].

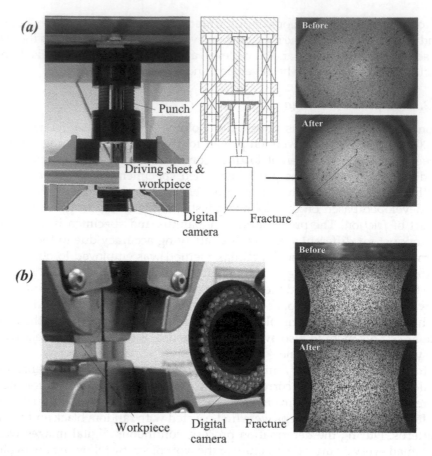

FIGURE 4.5
Experimental setups [13]: (a) Marciniak method and (b) Holmberg method.

In DIC measurement, a digital camera with a good resolution is generally used. The Marciniak experiments designed according to ISO 12004-2: 2008 with the miniaturization factor of 0.1 was conducted. The diameters of punch head, die, and the notched hole of driving plate were 10, 14, and 3 mm, respectively. The driving plate was made of annealed copper sheet metal with a thickness of 0.4 mm. The prepared plate was subjected to blanking process to fabricate the notched hole in the middle with satisfying edge quality so that the plate would not fracture before the specimen during the test.

4.2.2.2 Strain Measurement

A two-dimensional DIC processing program was developed to capture the displacement and further to calculate the strain of workpiece in deformation by using Matlab. In this DIC program, the open-source DIC program written by Eberl et al. [23] was used to generate the reference points at the deformation zone. Hence, the Matlab correlation function (cpcorr.m) in the image toolbox was modified for large displacement tracking. The full-field strain calculation approach, proposed based on the pointwise least-squares algorithm by Pan et al. [24], was then used to calculate the principal strains at each point.

To accurately determine the displacement and strain, the reference points array must be generated with a proper density. Too dense arrays could lead to a huge calculation work while too sparse points would reduce the accuracy. After many times of calculation tryout, the array of reference points with a separation distance of 15–20 pixels was found to be a satisfying solution. The array generated is shown in Figure 4.6.

As mentioned earlier, the tracking tool (cpcorr.m) provided within the Matlab software was then employed for DIC calculation. However, the original function of cpcorr.m abstracts the information of gray level within the square box, with the edge length of 10 pixels around the reference point. The square area with the edge length of 20 pixels in the predeforming image was then searched to find the best match pattern. Therefore, the displacement can be obtained after calculation. Nevertheless, the method is unsuitable for large displacement, considering that the program tends to underestimate the displacement by finding the match area in the predeformed image. The error can be ignored for small displacement. However, it could be accumulated for large displacement and multistep calculation. Therefore, the matching method, as shown in Figure 4.7, was employed by finding the best match of the chosen predeformed area in the image after deformation. In addition, the reference area was increased to 50 pixels for a better accuracy. The diagrams of the tracked results are shown in Figure 4.8.

To further calculate the strain field, a numerical differential algorithm is required. However, the direct differential method may amplify the displacement measurement error. Therefore, the pointwise least-squares algorithm proposed by Pan et al. [24] was employed to eliminate the amplification effect by introducing a smooth fitting solution. The calculated results are shown in Figure 4.9.

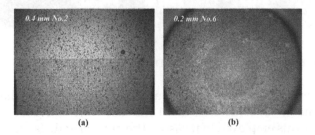

(a)   (b)

FIGURE 4.6
The tracking points matrix: (a) Holmberg test and (b) Marciniak test.

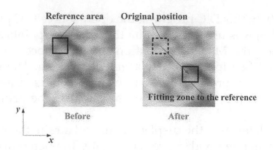

FIGURE 4.7
Schematic of the matching method.

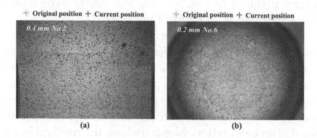

FIGURE 4.8
Tracking the points using DIC method: (a) Holmberg test and (b) Marciniak test.

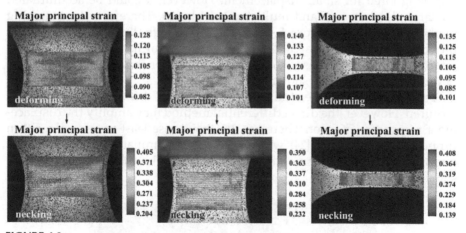

FIGURE 4.9

The strain distribution obtained based on the DIC method [68].

The accuracy of the DIC program was verified according to the tensile test of the specimen with a thickness of 0.4 mm, as shown in Figure 4.10a. A laser extensometer was employed to measure the deformed distance between two reflective marks. Meanwhile, the digital pictures of the deforming specimen were also taken, and the distance between the marks was calculated based

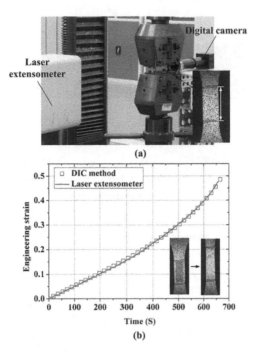

(a)

(b)

FIGURE 4.10
Verification of the accuracy of the DIC method [68]: (a) Tensile test with a laser extensometer and a digital camera and (b) Comparison of the results obtained by laser extensometer and DIC method.

on the DIC program. The engineering strains calculated based on the two methods are shown in Figure 4.10b.

4.2.2.3 Limit Strain Determination

After the strain field is obtained, the failure represented by necking and fracture and the limit strains can be further determined. As necking occurs, the deformation of material is localized at the necking area, leading to a sudden increase of strain. Based on these available data and information, the maximum major principal strain versus image sequence number curves are constructed to estimate the necking moment as illustrated in Figure 4.11. The maximum major strain increases steadily until the occurrence of local instability. The last image of the steady deformation is used to identify the initiation of necking. The limit strain is then obtained based on ISO 12004-2: 2008 and shown in Figure 4.12. The experiments under the same condition were conducted to verify the repeatability.

In addition, some researchers have revealed that the fracture limits the attainable deformation before necking happens in some deformation processes such as deep drawing for making the parts with complex structures

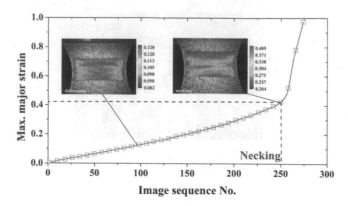

FIGURE 4.11
Determination of the necking moment [68].

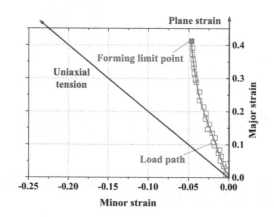

FIGURE 4.12
One point of the FLD and its load path [68].

and features [25] and the single-point incremental forming process [26]. Hence, the fracture FLD (FFLD) is attracting more attention [27,28]. To further characterize the size effect on the failure of sheet metals, the fracture limit strains were also identified. As shown in Figure 4.13, the major and minor strains just before the macroscopic fracture obtained by the DIC method were recorded as fracture limit strains.

4.2.3 Results and Discussion

After the forming limit experiments, the geometry and grain size effects on the forming limit were discussed. They are found to play a significant role in ductile failure in meso- and microforming processes. These effects were analyzed from four viewpoints:

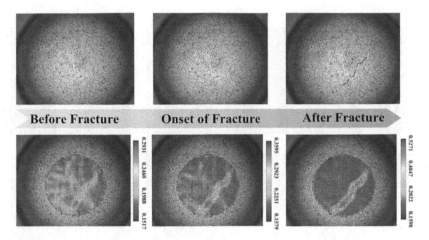

FIGURE 4.13
The strain distribution images onset of DF [69].

4.2.3.1 Deformation Process

The strain fields of deforming specimens were recorded and shown in Figure 4.14, which illustrates the effect of grain size on the deformation process.

In Figure 4.14a, the uniaxial tensile deformation of the specimens with a grain size of 27.5 µm and a thickness of 0.1 mm is demonstrated. The smooth and uniform distribution of strain in the measurement area is observed. With further deformation, strain localization gradually starts, leading to the necking and DF of the specimen. In contrast, Figure 4.14b shows the deformation of the specimen with a larger grain size of 90.7 µm and a thickness of 0.1 mm. The strain field is uneven from the beginning of the deformation. In addition, strain localization starts at the early stage of deformation. Unlike the smooth and uniform deformation, as shown in Figure 4.14a, the inhomogeneous deformation and irregular edge are obvious during the test. A similar observation was made during the Marciniak tests. Figure 4.14c shows the deformation of specimens with a grain size of 17.4 µm and a thickness of 0.2 mm. The strain field is uniformly distributed during the deformation process until the occurrence of necking. However, for the workpiece with a grain size of 166.2 µm, as shown in Figure 4.14d, the strain distribution is highly irregular, and an early strain localization was observed.

The different deformation behaviors of the specimens with different grain sizes were observed in the forming limit tests, which can be explained by the surface layer model. For the cases with small grain size, there are a number of grains over the thickness direction. The internal grains thus take up a major proportion. The overall material deformation tends to be uniform and smooth due to the complex interactive constraints and coordination of the individual grains with different orientations and properties.

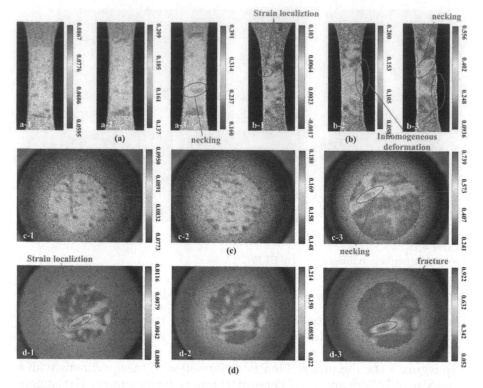

FIGURE 4.14
The plastic deformations of different specimens [13]: (a) $t = 0.1$, $d = 27.5$; (b) $t = 0.1$, $d = 90.7$; (c) $t = 0.2$, $d = 17.4$; and (d) $t = 0.2$, $d = 166.2$. (The units of t and d are mm and μm, respectively).

With the grain size approaching the thickness of sheet metal, the effect of surface grains becomes more evident. Since there are only a few grains in the deforming area, the orientation and property of individual grains thus have a significant influence on the overall deformation. The variable properties of different grains thus lead to an uneven deformation behavior and early strain localization.

4.2.3.2 FLD and FFLD Results

The FLD and FFLD results obtained from the forming limit tests are shown in Figure 4.15. The average necking and fracture limit strain for each type of specimen with Nos. 1–6 designed for the uniaxial to biaxial deformation conditions are shown in Figure 4.16.

Since the trend is similar for all six different specimens, for simplification, Nos. 1, 4, and 6 specimens were chosen as examples for discussion. From Figure 4.16, it can be found that the results of the specimens with a thickness of 0.4 mm show a satisfying repeatability. The FLC and fracture forming

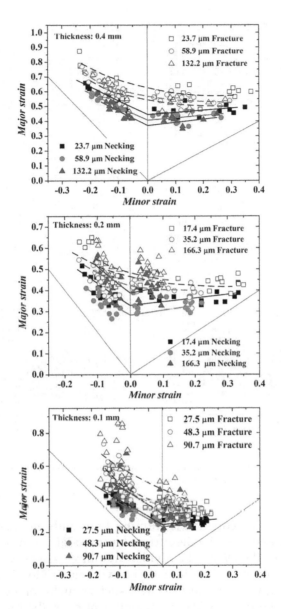

FIGURE 4.15
The forming limit results obtained [69]: (a) $t = 0.4\,mm$; (b) $t = 0.2\,mm$; and (c) $t = 0.1\,mm$.

limit curve (FFLC) shift down with the increase of grain size. In the figure, the major necking strains for the specimens Nos. 1, 4, and 6 are decreased from 0.641, 0.478, and 0.485 to 0.522, 0.421, and 0.408, respectively, with the reduction percentage of 10%–20% as the grain size increases from 23.7 to 132.2 μm. The fracture limit strains also demonstrate a similar deterioration tendency with the increase of grain size.

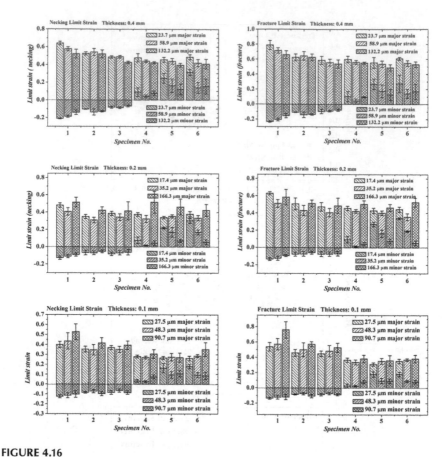

FIGURE 4.16
The average limit strains of the specimens with different thicknesses and grain szies [69]:
(a) $t = 0.4$ mm; (b) $t = 0.2$ mm; and (c) $t = 0.1$ mm.

The reason for the reduction of forming limit with the increase of grain size can also be explained by the surface layer model. As indicated by many research projects, the DF can be divided into three stages: void nucleation, growth, and coalescence. The initiation of void nucleation is driven by the stacking of dislocations at the flaw area, such as inclusion edges and grain boundaries. When the plastic deformation continues, the microvoids gradually grow until the occurrence of coalescence. The microcracks are then formed by the coalescence of microvoids and further lead to the fast loss of loading capacity and the occurrence of macroscopic fracture. Since the surface grains of the specimen are less constrained and their rotation facilitates the movement of dislocations, the void nucleation driven by the stacking and pileup of dislocations is thus less likely to occur at the surface layer. With the increase of grain size, the internal portion of the specimen becomes thinner. As a result, the coalescence of voids is easier to start, and the microcracks are also easier to grow in the thickness direction. In addition, the increase of grain size also makes the strain and void growth easier to be localized at the weak spots, leading to the reduction of limit strain.

With the increase of grain size from 17.4 to 35.2 µm for the specimens with a thickness of 0.2 mm, the forming limit of sheet metal is identified to decrease significantly, which is consistent with the observations of the specimens with a thickness of 0.4 mm. The necking major strains of specimen Nos. 1, 4, and 6 are found to decrease for 15.1%, 13.9%, and 11.6%, respectively. Nevertheless, with the increase of grain size to 166.3 µm, an abrupt variation of experimental result is observed, i.e., the repeatability of experiments becomes much worse and the scatter of the experimental results is significant. As shown in Figure 4.16b, the standard errors of the necking major strains for the specimen Nos. 1, 4, and 6 increase from 5.27%, 5.19%, and 8.38% to 10.9%, 24.4%, and 16.1%, respectively. Even though the formability of the materials seems to increase under this condition as the major limit strain increases, the decrease of the minor limit strain, however, shows a deviation of load path and an earlier local deformation condition.

Furthermore, a variation of the experimental results is also observed for the specimens with a thickness of 0.1 mm. With an increase of grain size from 27.5 to 48.3 µm, the scatters of the major necking strains for the specimen Nos. 1, 3, and 5 are increased from 8.00%, 5.51%, and 5.79% to 18.80%, 8.66%, and 13.60%, respectively. The tendency becomes more severe with the further increase of grain size to 90.7 µm when the corresponding scatters are increased to 14.6%, 10.9%, and 16.9%, respectively. This is because the DF is highly influenced by the individual grain properties when there are only one or few grains over the thickness direction. The fracture is thus determined by the deformation interaction of the several grains with different orientations and properties. The uneven distribution of the strain field and the early localization also contribute to the increase of uncertainties. Therefore, the scatter of the forming limit result becomes much worse, as revealed by the experimental results under that deformation condition.

4.2.3.3 Load Path

According to the observation of FLCs, the deviation of load path could be caused by an increase of grain size. To characterize this phenomenon, the $|\varepsilon_2/\varepsilon_1|$ ratio of the specimens with different geometries is shown in Figure 4.17. An FLC can be regarded as the line that links the forming limit points obtained based on the different load paths from uniaxial tension to biaxial stretching. During the deformation process, the load path represented by $\varepsilon_2/\varepsilon_1$ is close to constant until the occurrence of necking. Therefore, the variation of $\varepsilon_2/\varepsilon_1$ can be employed as an indicator to show the change of load path.

From Figure 4.17, a significant reduction of $|\varepsilon_2/\varepsilon_1|$ is observed with the increase of grain size, especially for the specimen Nos. 1 and 6, whose load paths are close to uniaxial and biaxial tensions, respectively. In addition, the

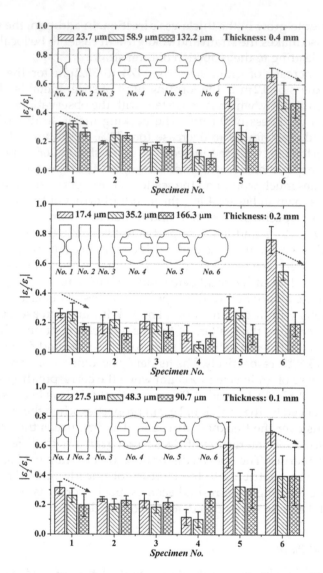

FIGURE 4.17
Size effect on the load paths [13]: (a) $t = 0.4$ mm; (b) $t = 0.2$ mm; and (c) $t = 0.1$ mm.

reduction is much more obvious with the decrease of t/d to two or less. For instance, the reductions of $|\varepsilon_2 / \varepsilon_1|$ for Nos. 1, 5, and 6 specimens with a thickness of 0.2 mm are −3.30%, 11.1%, and 27.9%, respectively, when the grain size is increased from 17.4 to 35.2 μm. Nevertheless, the reductions of $|\varepsilon_2 / \varepsilon_1|$ jump to 36.4%, 53.2%, and 64.2%, respectively, if the grain size is further increased from 35.2 to 166.3 μm. The $|\varepsilon_2 / \varepsilon_1|$ results of the specimens with a thickness of 0.1 mm also demonstrate a similar trend. With an increase of grain size, the reduction of $|\varepsilon_2 / \varepsilon_1|$ can reach 40%–50% for the specimen Nos. 5 and 6.

The deviation of load path is also associated with the change of deformation and failure behaviors when the grain size approaches the thickness of sheet metal. Since the surface grains have a dominant role in meso- and microforming of sheet metals, the early strain localization is observed due to the inhomogeneous deformation induced by the random nature of the individual grain's orientation and properties. In addition, the failure mode also experiences a significant change due to the effect of surface grains. The load path also changes when the grain size is increased to be close to the thickness of sheet metals.

4.2.3.4 Microstructure Observation

To find out the mechanism of the size effect on the failure behavior of sheet metals, the scanning electron microscope (SEM) observations of fracture surfaces were conducted. The SEM results of the specimens with a thickness of 0.4 mm are shown in Figure 4.18.

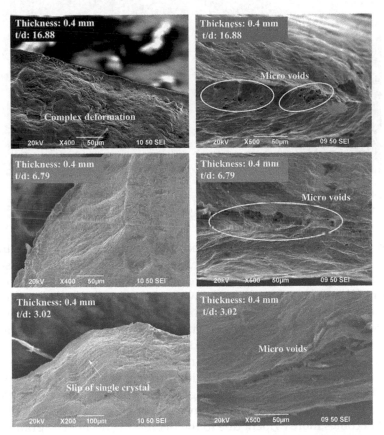

FIGURE 4.18
SEM observation of the specimens with a thickness of 0.4mm [69].

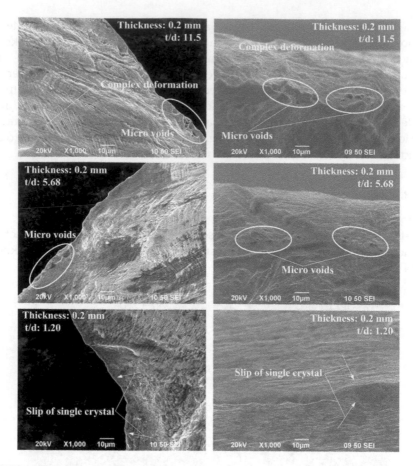

FIGURE 4.19
SEM observation of the specimens with a thickness of 0.2mm [69].

From the figure, microvoids can be easily observed at the fracture surfaces with different t/d conditions. The number of voids is found to decrease with an increase of grain size. The voids are also shown to gradually concentrate at the middle area of fracture surface with the increase of grain size. In addition, the complex plastic flow behaviors of materials are observed near the fracture surface. However, with an increase of grain size to 132.2 μm, the single crystal slipping can be observed as a contrast.

Figures 4.19 and 4.20 show the deformations of specimens with the thickness of 0.2 and 0.1 mm, respectively, with different values of t/d. The complex plastic deformations as well as the significant microvoids are observed for small grain size scenarios. With the increase of grain size, the slip of single crystal becomes more evident due to the decreasing number of grains in the thickness direction. The number of microvoids is also found to decrease due to the reducing proportion of the polycrystalline area, which is subjected

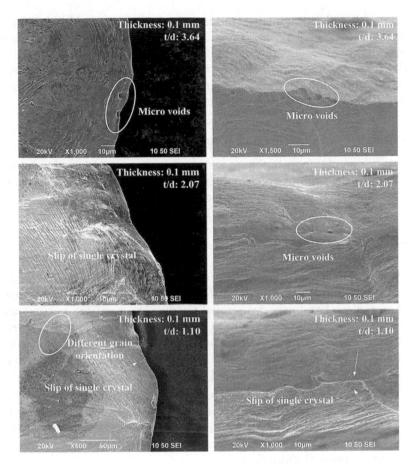

FIGURE 4.20
SEM observation of the specimens with a thickness of 0.1 mm [69].

to a complex interaction of grains. As t/d falls to two or less, there remains only one or two grains over the thickness direction. A knifelike edge is thus observed and the microvoids are seldom found.

The SEM observations verify the earlier discussion on the experimental results of forming limit and deformation behaviors. Since the surface grains are less constrained, their rotation facilitates the movement of dislocations. Therefore, void nucleation is less likely to occur due to the weak dislocation stacking behavior. As a result, the number of voids at the fracture surface is decreased with the flow stress of internal grains and the increase of grain size. In addition, the change of deformation and failure behaviors are linked with the single crystal slipping evidence according to the SEM observations, when the grain size approaches the thickness of sheet metal. When there are only one or two grains over the thickness direction, the grains are less constrained for more free surfaces. The material deformation thus tends to perform like a single crystal. With different orientations and properties of individual grains, the plastic deformation is easy to be localized at the weak spot, leading to significantly inhomogeneous deformation and early strain localization. This could also be the reason for the deviation of load path. Furthermore, the increase of uncertainty of individual grains also affects failure behavior. As a result, the results of the highly scattered forming limit are obtained.

4.2.4 Summary of Experimental Investigation

Many experimental investigations were conducted to explore the deformation and forming limit of sheet metals with different grain sizes and thicknesses. Different from the macroscopic results, the significant size effect on the deformation and failure behaviors can be observed as the grain size gradually approaches the thickness of sheet metal. The brief experimental findings are summarized in the following:

1. With the increase of grain size, the deformation behaviors in meso- and microforming gradually differ from those in macroscale, which has a more homogeneous strain distribution until the occurrence of necking. For the meso- and microscales, strain localization starts earlier during deformation, especially when there are only one or two grains over the thickness direction. An irregular deformation is also observed in the downscaled deformation tests.

2. The FLC is found to shift down with the increase of grain size, demonstrating the deterioration of formability of sheet metals. Highly scattered results are observed when the grain size approaches the thickness of sheet metal. The load path is also found to be affected by an increase of grain size.

3. According to the SEM observations, it shows that the fracture topography is strongly affected by the variation of geometry and grain sizes. With the increase of grain size, the number of microvoids is revealed to decrease evidently. When there are only one or two grains over the thickness direction, the microvoids can seldom be found, while the knifelike edges are observed. A significant regular single crystal slipping instead of the complex polycrystalline interactive deformation is observed with the decease of t/d to one or two.

4.3 Prediction of Size Effect Affected FLC Based on Theoretical Models

The prediction of FLC and modeling of failure behaviors of sheet metals have been a tantalized issue in recent years due to its wide application in industries. In fact, FLC can also be regarded as a validation and verification approach for various failure predictive methods of sheet metals due to its simplicity. So far, various failure theories have been well developed at macroscale and applied in the FLC prediction. As indicated by the experiments in meso- and microforming processes, on the other hand, the forming limit of sheet metals is highly affected by the size effect in the small-scaled forming scenario. It is thus important to know whether different macroscale models are applicable in microscale to establish the fundamental understanding of the failure mechanism at meso- and microscale. In this section, the applicability of different failure models and criteria is discussed and the obtained FLCs as described in the previous sections are analyzed. A study based on a damage-coupled model to describe how the size effect affects the DF behavior of sheet metal at micro- and mesoscale is also presented.

4.3.1 Swift Diffuse and Hill-Localized Criteria

The classical Swift [29] and Hill [30] criteria, which describe the diffuse and localized necking behavior of sheet metals, have been considered to be the most traditional method for the prediction of right and left sides of FLC, respectively.

According to the Swift criterion, diffuse necking starts when the external loads in the two principal directions stop increasing at the maximum level and the following condition is met:

$$\begin{cases} dF_1 = 0 \\ dF_2 = 0 \end{cases} \tag{4.7}$$

The following constraints can thus be derived:

$$\begin{cases} \dfrac{d\sigma_1}{\sigma_1} - dc_1 = 0 \\ \dfrac{d\sigma_2}{\sigma_2} - d\varepsilon_2 = 0 \end{cases} \tag{4.8}$$

σ_1 and σ_2 are the major and minor principal stresses, respectively. Similarly, ε_1 and ε_2 are the major and minor strains. The material is assumed to be isotropic and obeys the following flow rule:

$$\dot{\varepsilon}_i = \dot{\bar{\varepsilon}} \frac{\partial \bar{\sigma}}{\partial \sigma_i} \tag{4.9}$$

The effective stress increment $\dot{\bar{\sigma}}$ can be described as

$$\dot{\bar{\sigma}} = \frac{\partial \bar{\sigma}}{\partial \sigma_1} \dot{\sigma}_1 + \frac{\partial \bar{\sigma}}{\partial \sigma_2} \dot{\sigma}_2 = \frac{d\bar{\sigma}}{d\bar{\varepsilon}} \dot{\bar{\varepsilon}} \tag{4.10}$$

By combining Eqs. (4.8), (4.9), and (4.10), the Swift instability condition can be designated as

$$\frac{d\bar{\sigma}}{d\bar{\varepsilon}} = \left(\frac{\partial\bar{\sigma}}{\partial\sigma_1}\right)^2\sigma_1 + \left(\frac{\partial\bar{\sigma}}{\partial\sigma_2}\right)^2\sigma_2 \qquad (4.11)$$

Assuming that the material follows von Mises yield criterion and the effective stress and strain curve obeys the Hollomon hardening law of $\bar{\sigma} = K_h\bar{\varepsilon}^{n_h}$, the limit principal strain can then be derived as

$$\begin{cases} \varepsilon_1^{\text{limit}} = \dfrac{2n_h\left(1+\rho+\rho^2\right)}{(1+\rho)\left(2\rho^2-\rho+2\right)} \\[4mm] \rho = \dfrac{\varepsilon_2}{\varepsilon_1}, \qquad 0\le\rho\le1 \end{cases} \qquad (4.12)$$

On the other hand, the Hill criterion indicates that the through-thickness necking occurs while the deformation-induced hardening is equal to the softening effect caused by the reduction of thickness. The following constraints can be derived according to the Hill criterion:

$$\frac{d\bar{\sigma}}{\bar{\sigma}} = d\varepsilon_1 + d\varepsilon_2 \qquad (4.13)$$

By considering von Mises yield criterion and Hollomon hardening function, the limit strain is determined as

$$\begin{cases} \varepsilon_1^{\text{limit}} = \dfrac{n_h}{1+\rho} \\[4mm] -\dfrac{1}{2}\le\rho\le0 \end{cases} \qquad (4.14)$$

To verify the applicability of the earlier equation, the uniaxial tensile test results were further processed based on the Hollomon equation by using the least-squares method. The fitted results are shown in Figure 4.21 and Table 4.3.

The predicted FLCs based on the Swift and Hill criteria are shown in Figure 4.22. Since the highly scattered forming limit results are obtained for the specimens with a thickness of 0.1 and 0.2 mm with their maximum grain sizes, only smaller grain size conditions are thus discussed.

As shown in Figure 4.22, the predicted FLC agrees well with the experiments for the cases with a thickness of 0.4 mm and a grain size of 23.7 μm. However, the analytical results deviate from the experimental ones with the increase of grain size. For the specimens with the thickness of 0.1 and 0.2 mm, a significant overestimation of FLC is observed. This is because the Swift and Hill criteria are developed based on the homogeneous continuum

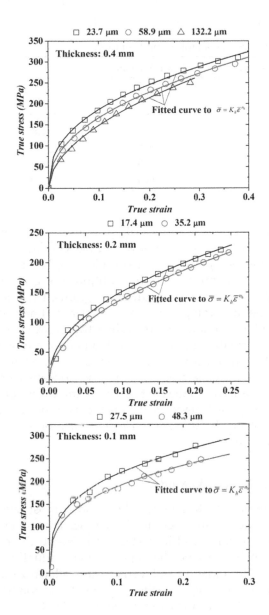

FIGURE 4.21
The fitted flow strain–stress curves according to $\bar{\sigma} = K_h \bar{\varepsilon}^{n_h}$ [69]: (a) 0.4 mm; (b) 0.2 mm; and (c) 0.1 mm.

hypothesis. According to the experimental investigation in the previous sections, it has been revealed that the influence of individual grains on the deformation and formability of materials becomes more significant at meso- and microscale. The failure behavior is highly correlated with the inhomogeneous deformation behaviors in forming process. The classical Swift and

TABLE 4.3

The Fitted Results of Hollomon Equation

Thickness/mm	Grain Size/μm	K	n
0.4	23.7	467.69	0.40
	58.9	469.57	0.45
	132.2	503.15	0.53
0.2	17.4	425.87	0.45
	35.2	429.22	0.49
0.1	27.5	439.81	0.31
	48.3	393.24	0.32

Hill criteria may not be able to describe the mechanism of the size effect described earlier. The complicated flow and necking behaviors induced by the interaction of several grains need to be addressed to deal with the failure behavior at meso- and microscale. In addition, the grain boundary area, which is the interactive surface of different grains, becomes more critical. At macroscale, the deforming grains are constrained and affected by each other leading to better homogeneity. When the grain size approaches the dimensions of deformation body, the development of damage becomes very different from that in macroscale deformation. The grain boundary area where dislocations are generally piled up and concentrated is believed to play a more important role in the DF formation and evolution of sheet metals at meso- and microscale. Therefore, the classical Swift and Hill criteria may not be able to describe the size effect affected forming limit of sheet metals in their original form.

4.3.2 Marciniak–Kuczynski Model

Marciniak–Kuczynski (M–K) model [31] is another most widely used model to analyze the instability in sheet metal forming. The model is based on an assumption that a geometrical imperfection exists in the deforming sheet metal, as shown in Figure 4.23. The imperfection gradually evolves with the increase of plastic deformation and leads to strain localization and necking in the end. In addition to its intuitive physical background and satisfying accuracy in different conditions, the model can also be easily incorporated with finite element (FE) analysis tools. The model has thus been widely investigated from different perspectives at macroscale. In this subsection, a simple and fundamental version of M–K model is employed to analyze the forming limit results obtained in the previous section and further validate the applicability of M–K model in the meso- and microforming scenarios.

As shown in Figure 4.23, $f_0^{M-K} = \dfrac{t_0^B}{t_0^A}$ is used to represent the initial ratio between the thicknesses of defective and flawless areas. In addition,

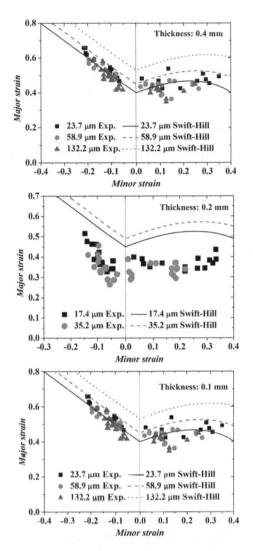

FIGURE 4.22
The predicted FLCs based on Swift and Hill criteria [69]: (a) 0.4 mm; (b) 0.2 mm; and (c) 0.1 mm.

the following conditions are met in the sheet metal forming to be discussed: (1) von Mises yield criterion; (2) plane-stress deformation condition $(\sigma_3 = 0)$; (3) geometrical compatibility $\left(d\varepsilon_2^A = d\varepsilon_2^B\right)$; (4) constant volume $(d\varepsilon_1 + d\varepsilon_2 + d\varepsilon_3 = 0)$; and (5) Levy–Mises flow rule. The following equation is then established based on the force equilibrium condition:

$$\sigma_1^A - \sigma_1^B f_0^{M-K}\left(e^{\varepsilon_3^B} - e^{\varepsilon_3^A}\right) = 0 \tag{4.15}$$

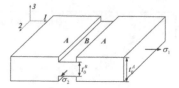

FIGURE 4.23
The schematic of M–K model [69].

Furthermore, the following relationships are derived based on previous assumptions:

$$
\begin{cases}
\dfrac{\varepsilon_2}{\varepsilon_1} = \dfrac{d\varepsilon_2}{d\varepsilon_1} = \rho \\[3mm]
\bar{\varepsilon} = \dfrac{2}{3}\sqrt{3}\varepsilon_1\sqrt{\rho^2 + \rho + 1} \\[3mm]
\eta^B = \dfrac{\sigma_2^B}{\sigma_1^B} = \dfrac{2\rho_B + 1}{2 + \rho_B} \\[3mm]
\varphi^B = \dfrac{\bar{\sigma}^B}{\sigma_1^B} = \sqrt{1 - \eta^B + \left(\eta^B\right)^2} \\[3mm]
\phi = \dfrac{\Delta\bar{\varepsilon}^B}{\Delta\varepsilon_1^B} = \dfrac{2\varphi^B}{2 - \eta^B} \\[3mm]
\dfrac{\rho + 2}{\sqrt{3\left(\rho^2 + \rho + 1\right)}}\bar{\sigma}^A - \dfrac{\bar{\sigma}^B}{\varphi^B}f_0^{M-K}\left(e^{\left.\varepsilon_3^B\right|_t + \frac{\Delta\bar{\varepsilon}^B}{\phi}(-1-\rho_B)} - e^{-\frac{\sqrt{3}}{2}\frac{\rho+1}{\sqrt{\rho^2+\rho+1}}\bar{\varepsilon}^A}\right) = 0
\end{cases}
\tag{4.16}
$$

Given the incremental major principal strain $\Delta\varepsilon_1^A$, $\Delta\varepsilon_1^B$ can be determined based on Eq. (4.16) by using Newton–Ranphson method. When $\Delta\varepsilon_1^B / \Delta\varepsilon_1^A \geq 7$ is satisfied, the strain localization at region B, as shown in Figure 4.23, is significant, and the major and minor strains of region A are considered to be the limit strains of necking. Otherwise, another increment of $\Delta\varepsilon_1^A$ is given, and the previous iteration calculation needs to be done again.

The fitted results of the equivalent stress and strain curves presented in Table 4.3 were employed to obtain the forming limits based on the M–K model. The experimental and analytical results are shown in Figure 4.24:

From Figure 4.24, it can be observed that the predicted FLCs agree with the experimental results under different geometry and grain size conditions. The analytical results were obtained by optimizing the value of f_0^{M-K}, which represents the deterioration of formability. However, the determination of f_0^{M-K} is the lack of theoretical basis. Actually, the M–K predictive result is sensitive to the value of initial defect. The determination of the parameter is

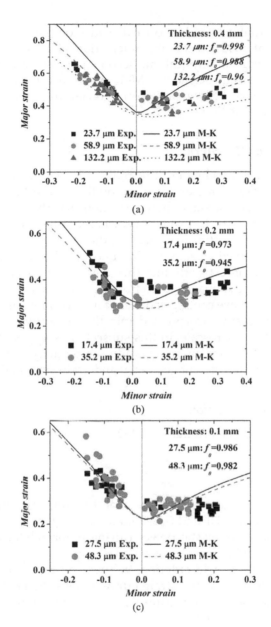

FIGURE 4.24
Prediction of the FLCs based on the M–K model [69]: (a) 0.4 mm; (b) 0.2 mm; and (c) 0.1 mm.

thus a critical issue that needs to be addressed. In meso- and microforming scenario, it has been shown that the deformation of individual grains is more significant, and the necking and strain concentration behaviors of the material are highly affected by the local interactive deformation of a few grains. Hence, during the application of the M–K model using different values of

f_0^{M-K}, f_0^{M-K} is actually extended to further include the internal defects, the effect of local microstructures, the interaction of individual grains, etc., even though its original physical implication is the surface flaw coefficient. Therefore, a further improvement of M–K model can be conducted in future to characterize the mechanism of size effect on failure behavior as well as the quantitative representation of f_0^{M-K}.

4.3.3 Uncoupled DF Criteria

DF criteria have been widely used in bulk and sheet metal forming. Since nonproportional and complex loading conditions are becoming popular, the application of DF criteria to these conditions has thus become a research focus in the recent years. Depending on whether the damage accumulation is considered in the constitutive equation, the DF criteria are classified into coupled and uncoupled categories. The uncoupled criteria do not include the damage evolution explicitly into the constitutive formulation in such a way the accumulated damage would not affect the yielding behaviors. As a result, they usually have concise and clear forms of stress and strain parameters and can be easily incorporated into the FE software. They are thus widely explored at macroscale. In this subsection, some of the widely used uncoupled DF criteria are discussed and used to analyze the experimental results obtained with different geometry and grain size conditions.

4.3.3.1 Freudenthal Criterion

Freudenthal criterion is formulated as [32]

$$\int_0^{\bar{\varepsilon}_f} \bar{\sigma} d\bar{\varepsilon} = C_1 \tag{4.17}$$

where $\bar{\sigma}$ and $\bar{\varepsilon}$ are the effective stress and strain, respectively. $\bar{\varepsilon}_f$ is the effective strain onset of DF and C_1 is a material constant. Freudenthal model actually calculates the plastic power during deformation. DF initiates as soon as the threshold power C_1 is reached.

Based on the Mises yield criterion, volume constancy, and the assumption that the equivalent stress–strain curve follows the Hollomon equation $\bar{\sigma} = K_h \bar{\varepsilon}^{n_h}$, the limit strains of fracture can be determined for the simple linear loading path $\rho = \varepsilon_2/\varepsilon_1 = $ constant as

$$\begin{cases} \varepsilon_{1f} = \sqrt{\dfrac{\dfrac{3}{4}\left(\dfrac{C_1(n_h+1)}{K_h}\right)^{\frac{2}{n_h+1}}}{\rho^2+\rho+1}} \\ \varepsilon_{2f} = \rho\varepsilon_{1f} \end{cases} \tag{4.18}$$

In Eq. (4.18), ε_{1f} and ε_{2f} are the limit major and minor strains, respectively. K_h and n_h are parameters of the Hollomon equation, which can be obtained according to the true stress–true strain curve. Hence, the FLC can be calculated for different loading paths by using different ρ values.

4.3.3.2 Ayada Criterion

Ayada et al. [33] included the hydrostatic stress $\sigma_m = \dfrac{\sigma_1 + \sigma_2 + \sigma_3}{3}$ into the DF criterion with the following format, and thus the so-called Ayada criterion has the following formulation:

$$\int_0^{\bar{\varepsilon}_f} \frac{\sigma_m}{\bar{\sigma}} d\bar{\varepsilon} = C_2 \tag{4.19}$$

where C_2 is a material constant, which determines the initiation of DF. Similar to the assumptions made during the derivation of forming limit results by using Freudenthal criterion, the following are obtained by using Ayada criterion:

$$\begin{cases} \varepsilon_{1f} = \dfrac{3C_2}{2(\rho+1)} \\ \varepsilon_{2f} = \rho\varepsilon_{1f} \end{cases} \tag{4.20}$$

The FLC based on this DF criterion can then be constructed by using different ρ values.

4.3.3.3 Cockcroft & Latham Criterion

The Cockcroft & Latham (C&L) criterion is designated in Eq. (4.21), which is focused on the effect of the maximum principal tensile stress on the failure behavior [34,35]:

$$\int_0^{\bar{\varepsilon}_f} \frac{\sigma_1}{\bar{\sigma}} d\bar{\varepsilon} = C_3 \tag{4.21}$$

The limit strain of fracture can be further formulated in the following based on C&L criterion:

$$\begin{cases} \varepsilon_{1f} = \dfrac{3C_3}{2(\rho+2)} \\ \varepsilon_{2f} = \rho\varepsilon_{1f} \end{cases} \tag{4.22}$$

4.3.3.4 Oyane Criterion

Oyane et al. [36] considered the growth of microvoids in plastic deformation and DF process and used the volume strain ε_v to estimate the occurrence of fracture. Oyane criterion is designated as follows:

$$\int_0^{\bar{\varepsilon}_f} \left(\frac{\sigma_m}{\bar{\sigma}} + C_4' \right) d\bar{\varepsilon} = C_4 \qquad (4.23)$$

The principal limit strains based on this criterion can be derived as

$$\begin{cases} \varepsilon_{1f} = \dfrac{3C_4}{2(\rho+1)+2\sqrt{3}C_4' \sqrt{\rho^2+\rho+1}} \\ \varepsilon_{2f} = \rho\varepsilon_{1f} \end{cases} \qquad (4.24)$$

The four criteria mentioned earlier are widely used to construct the FFLD at different geometry and grain size conditions. Different from the Swift/Hill criteria and M–K model, which describe the onset of necking, the DF criteria estimate the condition of fracture, which occurs after necking with the deformation progress localized at the necking region. As discussed in Section 4.2.3.2, both FLCs (onset of necking) and FFLCs (just before DF) were obtained during the experiments. To verify the four criteria in this section, the FFLCs were then employed considering these criteria developed for DF prediction. By using the least-squares method, the coefficients in the DF criteria can be determined by fitting the experimental data. Figure 4.25 shows both the experimental and analytical results.

Furthermore, it can be observed that the Freudenthal, C&L, and Ayada criteria can describe the FFLD with an acceptable accuracy. The errors between the experimental and analytical results are different for different loading conditions. For example, all the three criteria tend to underestimate the fracture limit strains for the bistretching condition. Oyane criterion, on the other hand, shows a satisfying description of the limit strain on the left side of FFLD. The analytical results for the right side of FFLD based on this criterion, however, deviate from the experimental ones significantly.

According to the application of different uncoupled DF criteria, it is found that these models are able to describe the experimental FFLCs of specimens obtained at different thickness and grain size conditions. However, it must be mentioned that the simplified calculation procedure introduced as earlier is not accurate, since the load path is assumed to be constant during the forming process. Nevertheless, this assumption is not true since necking usually occurs before DF, which would lead to a deviation of load path to plane strain. To realize more accurate prediction, the FE simulation can be an efficient way in the future. It should also be mentioned that the deforming-necking behavior of material is not influenced by introducing these criteria, since the uncoupled DF criteria did not include the effect of internal damage on the

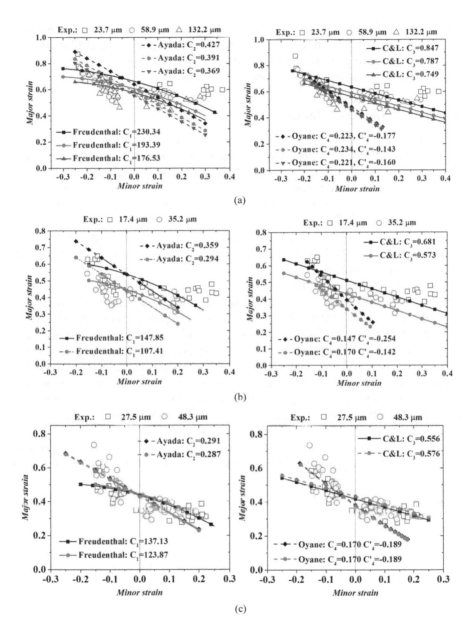

FIGURE 4.25
Prediction of the FLCs based on DF criteria [69]: (a) 0.4 mm; (b) 0.2 mm; and (c) 0.1 mm.

material deformation behavior explicitly. Hence, the size effect of FLC at neck-ing could not be well addressed by using these criteria. In addition, there is a lack of description and characterization of the intrinsic mechanisms of form-ing limit reduction at meso- and microscales. Therefore, to characterize the as-observed size effect of FLC and DF behavior of sheet metals at meso- and microscale, finding a convincing explanation of the mechanism of size effect on the DF process at a small scale and development of an efficient DF modeling approach to represent the whole deforming–necking–fracturing behavior is a challenging and eluded issue to be addressed.

4.3.4 GTN–Thomason Model

In general, the plastic deformation and DF process of sheet metals can be regarded as a three-step progress, as shown in Figure 4.26. When the plastic deformation starts, defects such as particles, inclusions, and other dislocation concentrating areas in metal alloys provide initiation sites for microvoids to nucleate. With the increase of plastic deformation, the nucleation of microvoids continues, and the volume of existing voids also grows simultaneously. When the volume of adjacent voids reaches a critical threshold, void coalescence will start, leading to a quick loss of load-carrying capacity and soon the macroscopic fracture happens.

Based on the void evolution process, the coupled DF criteria are proposed, which incorporate the damage accumulation into the constitutive equation to describe the whole deforming–necking–fracturing progress. Compared with the uncoupled criteria, the coupled ones provide a better and more accurate description of DF in deformation process. In addition, many important factors and aspects such as void distribution, geometry, dimension, etc. on the deformation and DF behaviors that are ignored or simplified in uncoupled DF criteria can be included and well treated. Therefore, more attention has been paid to the coupled criteria and their improvement and application in sheet metal forming scenario in recent years.

Gurson criterion [37] is one of the most attractive coupled DF models. By considering the growth of microvoids in a volume-limited cell, Gurson

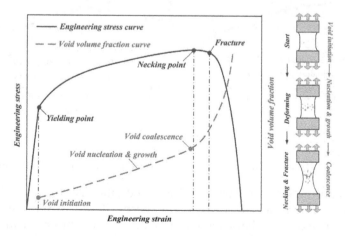

FIGURE 4.26
A schematic diagram of plastic deformation and void evolution process.

criterion provides a realistic description of material microstructure in DF process and associates DF to micromechanical parameters directly. In addition, Tvergaard and Needleman et al. [38–40] extended the Gurson criterion by including the isotropic hardening and void nucleation in such a way that the criterion is then formulated and addressed as the well-known Gurson–Tvergaard–Needleman (GTN) model. The GTN model has been widely explored for different aspects. Among these efforts, Besson [41] discussed the kinematic hardening and the viscoplastic deformation mechanism based on the GTN model. Gologanu et al. [42] analyzed the effect of void shape under different loading conditions. Wen et al. [43] and Monchiet et al. [44] discussed the void size effect by introducing the strain gradient theory into the model.

Furthermore, many researchers also put efforts on the study of coalescence of voids, since it is directly related to the onset of DF process. Experimentally, Weck et al. [45] and Hosokawa et al. [46] investigated the void growth and coalescence by preparing the embedded array of holes after diffusion bonding of two sheets filled with laser-drilled holes. By employing an in situ X-ray computed tomography, the growth and coalescence of these preimplanted holes can be clearly observed. The validity and efficiency of different criteria in analysis of void growth and coalescence are thus validated. They have developed a new approach for investigating the DF and void evolution behavior, since it is rather difficult to capture the void coalescence experimentally, let alone to investigate the in situ evolution process. By employing the FE simulation, Liu et al. [47] discussed the coalescence of voids under various loading conditions. Zhang et al. [48] analyzed the effect of stress triaxiality on void coalescence. In addition, by using the atomistic and molecular simulations, Rudd et al. [49] investigated the void growth and coalescence in polycrystalline copper. Potirniche et al. [50] explored the void evolution in single crystal material. In addition to experimental research, modeling has also been the focus of study. The pioneering work on this issue was made by Thomason [51–53]. By analyzing the internal necking behavior in the alignment of microvoids, they proposed a novel model to capture the onset of coalescence in an intuitive manner. Thomason model has been well accepted and validated by simulations as well as experiments. On the basis of this model, a lot of improvements have been made. Among them, Pardoen et al. [54] considered the effect of material hardening based on the original Thomason model. Scheyvaerts et al. [55] included the direction of local deforming surface during the internal necking in the alignment of microvoids. Fabregue et al. [56] analyzed the effect of second-phase particles on the overall coalescence behavior. Zhang et al. [57] and Pardoen et al. [54] also studied the feasibility of predicting material failure by combining Gurson model with Thomason model. In such a way, a complete process of void nucleation, growth, and coalescence can be explicitly well described.

Many efforts have also been provided to explore the applicability of the Gurson-based models. The Gurson-based models are usually incorporated into FE codes to simulate various plastic-deforming processes and further to verify their applicability. Among these articles, Oh et al. [58] explored the fracture of API X65 steel tensile bars, Yan et al. [59] studied the edge cracking of silicon steel, Li et al. [60] conducted the tensile and compression tests of various Al-alloy 6061 specimens, Abbasi et al. [61] and Uthaisangsuk et al. [18,62] explored the forming limit of various sheet metals. The simulation results are found to have a good agreement with the experimental ones.

So far, modeling and investigation of damage evolution in material deformation process is still an eluded and tantalized issue for many researchers due to its association with the formability of various metal materials. However, the exploration on the damage and fracture in meso- and microforming has not been well conducted. In this section, the GTN–Thomason model is employed to predict the FLD of sheet metals with different geometry and grain size conditions.

The GTN–Thomason model is formulated by three submodels that describe the nucleation, growth, and coalescence of microvoids, respectively. The yield criterion of GTN model describing the void growth with deformation is as follows [39]:

$$\Phi = \left(\frac{\sigma_v}{\sigma_y}\right)^2 + 2q_1 f^* \cosh\left(-\frac{3}{2}\frac{q_2\sigma_m}{\sigma_y}\right) - \left(1 + q_3 f^{*2}\right) = 0 \qquad (4.25)$$

where $\sigma_v = \sqrt{\frac{3}{2}S_{ij}S_{ij}}$ is the von Mises equivalent stress, $\sigma_m = \frac{1}{3}\sigma_{kk}$ is the hydrostatic stress, $S_{ij} = \sigma_{ij} - \sigma_m\delta_{ij}$ is the deviatoric component of Cauchy stress, δ_{ij} is the Kronecker delta, σ_y is the equivalent stress of the base material in which the voids exist, and q_1, q_2, and q_3 are the coefficients introduced. In addition, f^* is the effective void volume fraction used to describe the quick loss of stress-carrying capacity after void coalescence and is formulated as

$$f^* = \begin{cases} f & f < f_c \\ f_c + \dfrac{f_u - f_c}{f_f - f_c}(f - f_c) & f \geq f_c \end{cases} \qquad (4.26)$$

In Eq. (4.26), f_c is the void volume fraction at the onset of coalescence. $f_u = 1/q_1$ is the void volume fraction when the load-carrying capability of material equals zero. f_f is the final void volume fraction when failure occurs.

In such a way, the connection between the void volume fraction f and the plastic deformation process is thus established. In addition to that, the evolution of f should also be constructed. The increase of void volume fraction $\dot{f}$ that is contributed by void growth and nucleation can be formulated as follows:

$$\dot{f} = \dot{f}_{\text{growth}} + \dot{f}_{\text{nucleation}} \qquad (4.27)$$

The growth of existing voids can be represented as

$$\dot{f}_{growth} = (1-f)\dot{\varepsilon}^p_{kk} \tag{4.28}$$

where $\dot{\varepsilon}^p_{kk}$ is the hydrostatic component of the plastic strain rate.

To characterize the void nucleation process, the well-known strain-controlled void nucleation model of Needleman [40] is employed as follows:

$$\begin{cases} \dot{f}_{nucleation} = A\dot{\bar{\varepsilon}}^p_m \\ A = \dfrac{f_n}{s_N\sqrt{2\pi}}\exp\left(-\dfrac{1}{2}\left(\dfrac{\bar{\varepsilon}^p_m - \varepsilon_N}{s_N}\right)^2\right) \end{cases} \tag{4.29}$$

As shown in Eq. (4.29), the nucleation rate is assumed to follow a normal distribution of the equivalent plastic strain $\bar{\varepsilon}^p_m$ of the base material. f_n is the nucleation speed coefficient, s_N and ε_N are the standard deviation and the mean value of the normal distribution, respectively.

Therefore, the void evolution in damage formation process is well formulated. However, the void coalescence is only represented by the critical indicator f_c. Since the critical porosity f_c cannot be considered a material constant as it is related to various factors such as void spacing, void shape, stress and strain state, etc. [54,57], the Thomason model [51] is thus employed to describe the coalescence of voids in the following:

$$\begin{cases} \dfrac{\sigma_{zz}}{\sigma_y} = (1-\chi^2)\left[0.1\left(\dfrac{1}{\chi}-1\right)^2 + 1.2\sqrt{\dfrac{1}{\chi}}\right] \\ \chi = \left(\dfrac{3}{2}f\lambda_0\left(\dfrac{3}{2}e_{zz}\right)\right)^{1/3} \end{cases} \tag{4.30}$$

where σ_{zz} is the major principal stress applied on the unit cell. e_{zz} is the major principal strain. λ_0 is a fitted coefficient denoting the spatial distribution of microvoids. Based on Besson's research [41], void coalescence starts when χ reaches a critical value of $\chi_c = 0.9$. When χ is increased to χ_c, the corresponding value of f can be regarded as the critical indicator f_c.

The GTN–Thomason model can be implemented into ABAQUS/Explicit via the user material subroutine VUMAT [63]. To realize the accurate description of material behavior, the parameters of the established model need to be determined. According to previous research [13,64,65], some of these parameters are given in Table 4.4. On the other hand, the other two parameters f_n and λ_0 can be obtained by finding the coincidence of the simulation and experimental results of uniaxial tensile curves according to Abassi et al. [61].

TABLE 4.4

The Parameters of the GTN–Thomason Model

Parameters	ε_N	s_N	q_1	q_2	q_3	κ	χ_c	$f_f - f_c$
Value	0.3/0.2	0.1	1.5	1	2.25	1	0.9	0.15

$\varepsilon_N = 0.3$ for 0.4-mm thick specimens, $\varepsilon_N = 0.2$ for 0.2- and 0.1-mm thick specimens.

TABLE 4.5

The Fitted Values of f_n and λ_0

Thickness (mm)	Grain size (µm)	f_n	λ_0
0.4	23.7	0.0314	7.1
	58.9	0.0285	10.1
	132.2	0.0265	16.2
0.2	17.4	0.038	9.6
	35.2	0.023	43
0.1	27.5	0.023	15.5
	48.3	0.0157	29.5

Hence, the engineering stress–strain curve is used as an objective function in optimal determination of the two parameters. The highest point of the experimental curve is first employed to obtain the value of f_n. The highest point is usually regarded as the starting of diffusive instability. Since the beginning of void coalescence would lead to an immediate drop of load-carrying capacity, it is thus assumed that the coalescence has not started at the diffusive instability point. The flow stress at the highest point is considered to be solely related to f_n, considering that λ_0 is a coefficient in the void coalescence process. After f_n is determined, λ_0 can be obtained by minimizing the difference between the simulated and experimental results at the failure point when a sudden drop of the strain–stress curve takes place, representing the fracture occurrence after void coalescence. The calculated parameters for different geometry and grain size conditions are shown in Table 4.5. The simulated and experimental engineering stress–strain curves are shown in Figure 4.27. A good agreement between the FE and experimental results can be observed.

After all the parameters were obtained, the FE model, as shown in Figure 4.28, was established to realize different loading paths by applying different displacement conditions in x and y directions. Based on the major and minor strain history data of the necking spot, the necking FLCs were constructed according to ISO 12004-2: 2008. The FFLCs were also recorded at the onset of $f = f_f$. In addition, the middle FLCs satisfying the criterion of $f = f_c + (f_f - f_c)/2$ were also obtained. The predicted results are compared with the experimental ones and are shown in Figures 4.29–4.31.

Both FLC and FFLC obtained show agreement with the experimental results. By using the parameters given in Table 4.5 at different geometries and

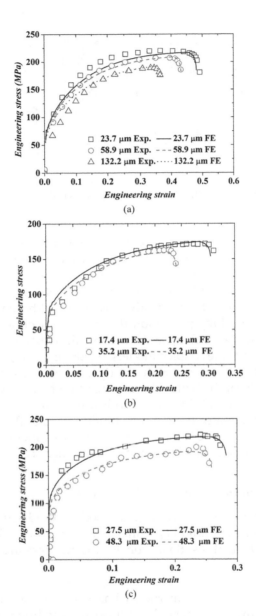

FIGURE 4.27
Comparison of the simulated and experimental engineering stress–strain curves [69]: (a) 0.4 mm; (b) 0.2 mm; and (c) 0.1 mm.

grain sizes, the reduction of forming limit with the increase of grain size can be described by introducing different parameter values. The deterioration of formability with the increase of grain size in the simulation was realized by increasing the value of λ_0, which actually represents the difficulty of coalescence. As discussed in the experimental section, the reduction of formability

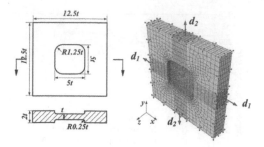

FIGURE 4.28
The FE model established to realize different loading paths [69].

with the increase of grain size could be attributed to the increasing coalescence with the decrease of void spreading area and the increasing trend of strain localization. Therefore, the GTN–Thomason model has the potential to describe how the size effect affects the failure behavior of sheet metals at meso- and microscale. This is mostly because GTN-based models are different from other necking-prediction theories or uncoupled DF criteria. In addition, to predict the necking and fracture failure behavior under various loading conditions, these coupled models are also excelling in characterizing the damage evolution process with clearly defined coefficients. Hence, the mechanism of DF can be better characterized with stronger physical background. Even though the GTN-based models are complex in formula and the accurate determination of their many coefficients is still under discussion, these models have been one of the most attractive methods in the recent development of sheet metal forming limit prediction technologies.

4.3.5 Summary of the Theoretical Discussion

Various macroscale failure theories and models are discussed in this section to validate their applicability in meso- and microscale for forming limit prediction and DF analysis. Some concluding remarks are summarized in the following:

1. The analytical FLCs constructed based on the Swift/Hill criteria agree with the experimental results at the conditions close to macroscale (such as the specimen with thickness of 0.4 mm and the smallest grain size). Nevertheless, the analytical results deviate from the experimental ones when the grain size approaches the geometry dimension. Since the Swift/Hill model is derived based on the assumptions of homogeneity and continuum, the increasing tendency of the localized deformation with the increase of grain size, as revealed in the experiments, needs to be considered for a better description of meso- and microforming process.

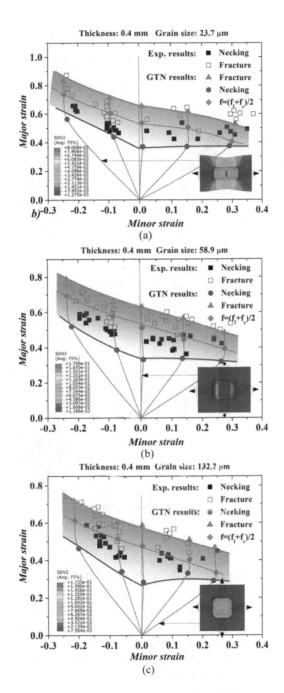

FIGURE 4.29
The predicted forming limit results of the specimens with a thickness of 0.4 mm based on the GTN–Thomason model [69]: (a) $d = 23.7\,\mu m$; (b) $d = 58.9\,\mu m$; and (c) $d = 132.2\,\mu m$.

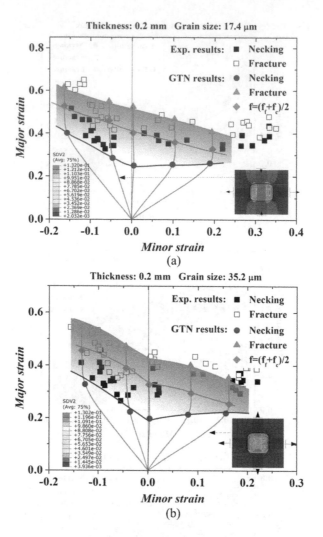

FIGURE 4.30
The predicted forming limit results of the specimens with a thickness of 0.2 mm based on the GTN–Thomason model [69]: (a) $d = 17.4\,\mu m$ and (b) $d = 35.2\,\mu m$.

2. The M–K method is able to model the reduction of formability of materials with the increase of grain size for the specimens with different thicknesses by adjusting the coefficient f_0^{M-K} of initial surface damage. However, the parameter-determination method still needs further discussion for its successful application in meso- and microforming processes.

3. Uncoupled DF criteria have been employed to determine the FFLCs. Similar to the M–K model, the damage coefficients of the DF criteria are optimized via minimizing the difference between the analytical

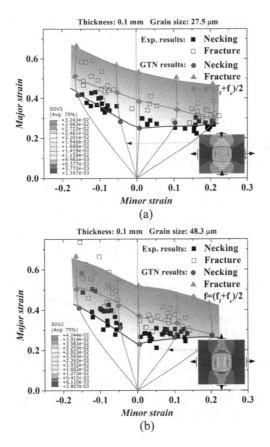

FIGURE 4.31
The predicted forming limit results of the specimens with a thickness of 0.1 mm based on the GTN–Thomason model [69]: (a) $d = 2.7.5\,\mu m$ and (b) $d = 48.3\,\mu m$.

and experimental results. In addition, the used criteria presented in this book tend to underestimate the forming limit at the biaxial stretching state.

4. The coupled GTN–Thomason model is also discussed for determining the forming limit of materials with different geometry and grain size conditions. The model is found to be able to provide a satisfying estimation of both FLC and FFLC at different meso- and microscales. In addition, the coefficient variation at different conditions is also found to reflect the mechanism of size effect on the increasing coalescence behavior in the forming at meso- and microscale. The GTN–Thomason model itself, however, still lacks the capability for reflecting the size effect on ductile damage and fracture in deformation process. The improvement and modification are required for its

successful application in the small scale to describe the mechanism of size effect on the damage and fracture evolution in deformation process.

4.4 Modeling of Size Effect Based on the GTN–Thomason Model

Considering the advantages of the GTN–Thomason model, such as the intuitive description of damage evolution and the good capability of describing both the necking and fracture, the modification of the model to include the size effect on DF behavior to represent how the scale factor affect the failure behavior of sheet metals at meso- and microforming processes is critical as a further step.

In tandem with this goal, the GTN–Thomason model was extended by incorporating the surface layer model to describe the effect of scale factor, i.e. t/d, on the void nucleation, growth, and coalescence, and the detailed realization procedure is presented in Figure 4.32. The extended model was then employed in the FE simulation of forming limit tests. The strain–stress curves of the uniaxial stretch were then obtained and compared with the simulation results to determine the damage and size-related parameters. The forming limits with different scale factors were predicted and compared with the FLCs obtained by experiment.

4.4.1 Modeling Process Considering Size Effect

4.4.1.1 Void Nucleation

As illustrated in Section 4.3.4, the strain-controlled void nucleation model proposed by Needleman [40] can be designated as follows:

$$\begin{cases} \dot{f}_{\text{nucleation}} = A\dot{\bar{\varepsilon}}_m^p \\ A = \dfrac{f_n}{s_N\sqrt{2\pi}}\exp\left(-\dfrac{1}{2}\left(\dfrac{\bar{\varepsilon}_m^p - \varepsilon_N}{s_N}\right)^2\right) \end{cases} \tag{4.31}$$

As indicated by the experimental investigation, the nucleation of voids is affected by the grain and geometry sizes significantly. This is because microvoids are mainly nucleate at particles, inclusions, and grain boundaries that obstacle dislocation motion and lead to stress concentration [3,49]. According

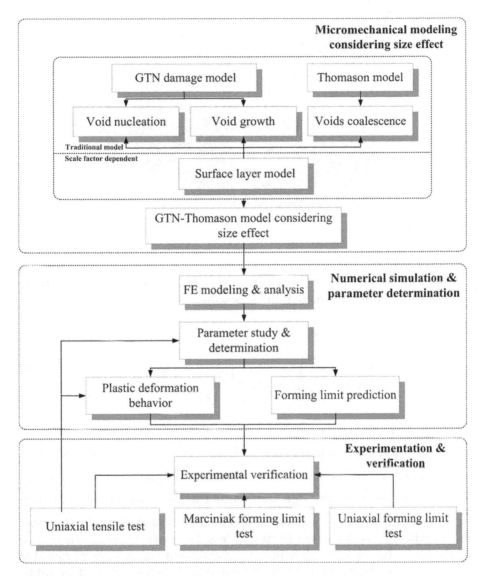

FIGURE 4.32
The modeling procedure of the extended GTN–Thomason model [13].

to the surface layer model, the surface grains are less constrained for their free surfaces. As a result, their rotation facilitates dislocation movement. Thus, the void nucleation seldom takes place in the surface grains. With the reduction of t/d, the proportion of surface grains is increased, leading to a significant reduction of "void nucleation band," as shown in Figure 4.33.

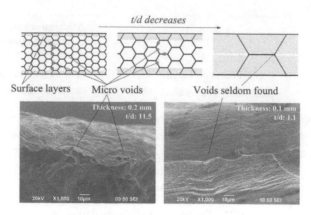

FIGURE 4.33
The size effects of void nucleation [13].

To consider the size effect of void nucleation, Eq. (4.31) is extended with the scale coefficient of t/d shown in the following:

$$
\begin{cases}
\dot{f}_{\text{nucleation}} = A\dot{\bar{\varepsilon}}_m^p \\[2ex]
A = \dfrac{\left(1-\dfrac{d}{t}\right)f_n}{s_N\sqrt{2\pi}}\exp\left(-\dfrac{1}{2}\left(\dfrac{\bar{\varepsilon}_m^p-\varepsilon_N}{s_N}\right)^2\right), \quad \dfrac{t}{d}>2
\end{cases}
\tag{4.32}
$$

In macroscale, the grain size is much smaller than the thickness of sheet metals, $\dfrac{d}{t}\approx 0$ and the original void nucleation model can be obtained based on Eq. (4.32). When the grain size approaches the geometrical dimensions of sheet metals, $\dfrac{d}{t}$ becomes greater, leading to the reduction of nucleation speed coefficient A and the decreasing tendency of nucleation. On the other hand, when t/d is decreased to two or less, i.e., there are only one or two grains over the thickness direction, the microvoids can seldom be found at the fracture surfaces and the fracture behavior is highly affected by single crystal deformation as revealed in the research conducted by Fu et al. [3] and Furushima et al. [5] on the observation of microvoids at the fracture surface. Since the void evolution is not significant, the GTN model may no longer be well applicable in this scenario. Therefore, the condition with t/d less than two is not discussed in the present modeling process.

4.4.1.2 Void Growth

When the dimensions of workpieces are decreased from macro to meso- and microscale, the decrease of flow stress has been presented and explained thoroughly in the previous chapters. Obviously, this effect affects the deformation behavior of materials in void growth and should be characterized in micromechanical modeling process with different scales. To model the decrease of flow stress, the surface layer model, which has been introduced in the previous chapters, was used to describe the size effect of base material. The extended GTN model is thus formulated as

$$
\begin{cases}
\Phi = \left(\dfrac{\sigma_v}{\sigma_y}\right)^2 + 2q_1 f^* \cosh\left(-\dfrac{3}{2}\dfrac{q_2 \sigma_m}{\sigma_y}\right) - \left(1 + q_3 f^{*2}\right) = 0 \\[2mm]
\sigma_y\left(\bar{\varepsilon}_y\right) = \bar{\sigma}_y^{\text{ind}}\left(\bar{\varepsilon}_y\right) - \bar{\sigma}_y^{\text{dep}}\left(\bar{\varepsilon}_y\right) \\[2mm]
\bar{\sigma}_y^{\text{ind}}\left(\bar{\varepsilon}_y\right) = M\tau_R\left(\bar{\varepsilon}_y\right) + \dfrac{k\left(\bar{\varepsilon}_y\right)}{\sqrt{d}} \\[2mm]
\bar{\sigma}_y^{\text{dep}}\left(\bar{\varepsilon}_y\right) = \eta\left(M\tau_R\left(\bar{\varepsilon}_y\right) + \dfrac{k\left(\bar{\varepsilon}_y\right)}{\sqrt{d}} - m\tau_R\left(\bar{\varepsilon}_y\right) \right)
\end{cases}
\tag{4.33}
$$

In this equation, the equivalent stress of based material σ_y in the original GTN model is represented based on the surface layer model to include the reduction of flow stress caused by the grain size effect. As discussed in Chapter 2, σ_y is constructed by size-dependent and -independent terms of $\bar{\sigma}_y^{\text{dep}}$ and $\bar{\sigma}_y^{\text{ind}}$, respectively. M is the orientation factor regarding the slips on deformation systems. m is the orientation factor in a single crystal. τ_R is the critical shear resolved stress of the single crystal. $k\left(\bar{\varepsilon}_y\right)$ is the locally intensified stress required to propagate general yield across the polycrystalline grain boundaries. The details of $\bar{\sigma}_y^{\text{dep}}$ and $\bar{\sigma}_y^{\text{ind}}$ can be found in Chapter 2.

It should be noticed that a simplification assumption is introduced that the equivalent stress–strain relationships obtained in the uniaxial tensile tests are regarded as the constitutive behavior of the base material directly. The softening effect on the flow stress induced by the existence of microvoids is ignored. Actually, it is difficult to obtain the constitutive parameters of the base material directly by experiments.

4.4.1.3 Void Coalescence

The Thomason model is employed to model void coalescence. The model assumes that coalescence starts when the stress concentration occurs in the intervoid ligaments. For simplification, the material is assumed to be formulated by regular cylindrical cells with a height of $2H$ and a width of $2L$, containing a spherical void with a diameter of $2R$, as shown in Figure 4.34.

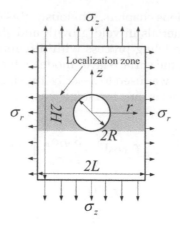

FIGURE 4.34
Thomason model of spherical voids [13].

Based on the schematics of the model, the Thomason model is formulated as [51,52]

$$\begin{cases} \dfrac{\sigma_{zz}}{\bar{\sigma}_m} = (1-\chi^2)\left[0.1\left(\dfrac{1}{\chi}-1\right)^2 + 1.2\sqrt{\dfrac{1}{\chi}} \right] \\ \chi = \dfrac{R}{L} \end{cases} \tag{4.34}$$

where σ_{zz} is the major principal stress applied on the unit cell and $\bar{\sigma}_m$ is the flow stress in the ligament. According to Benzerga et al. [66] and Besson [41], χ is represented as

$$\chi = \left(\frac{3}{2} f\lambda_0 \left(\frac{3}{2}\kappa e_{zz} \right) \right)^{1/3} \tag{4.35}$$

where e_{zz} is the major principal strain and κ is a fitting parameter used to represent the irregularity since the actual material is not a perfect array of representative cells. κ is taken as 1. λ_0 is the initial value of H/L, denoting the spatial distribution of microvoids.

As demonstrated by experiments, the coalescence of voids is affected by size effect. The void spreading area in the thickness direction of sheet metal is decreased due to the increasing proportion of surface layer when the grain size approaches the sheet thickness. As a result, the initiation and evolution of void coalescence is facilitated and leads to the earlier failure of sheet metals. In addition, the inhomogeneity of material increases with grain size in meso- and microscale. Therefore, the strain localization at the weak spots is also likely to occur and contribute to the earlier void coalescence.

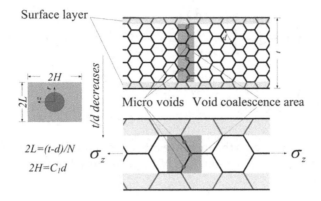

FIGURE 4.35
The effect of geometry and grain sizes on void distribution [13].

To include size effect in coalescence analysis, the surface layer model is included and shown in Figure 4.35. The microvoids are most likely to be formed at grain boundaries [3,49]; therefore, the distribution of voids is highly related to grain and geometry sizes. The factor λ_0, which represents the spatial distribution of voids, thus needs to be introduced to consider the size effect.

Due to $\lambda_0 = H/L$, the height H and width L of the representative cell are considered via λ_0. Obviously, the thickness is much smaller than the other two dimensions in sheet metal forming process, and fracture always occurs across the thickness direction. Void coalescence is most likely to occur in the thickness direction of sheet metal. Therefore, it is assumed that L is in the thickness direction and H is in the direction of principal stress. With the increase of grain size, the grain boundary density and the number of voids are decreased and shown in Figure 4.35. The height H of the representative cell is thus increased with grain size. And H is assumed to be directly proportional to the average diameter of grains and denoted as

$$2H = C_1 d \tag{4.36}$$

On the other hand, there are much fewer grains over the thickness. The void spreading is affected by the existence of surface layer as discussed earlier. Hence, L cannot be regarded to be in direct connection with the grain size. To consider the effect of surface layer, L is represented by the ratio between the thickness of the internal layer with the average number N of voids over the thickness as follows:

$$2L = \frac{t-d}{N} \tag{4.37}$$

where N is assumed to be related to the nucleation and designated as

$$
\left\{
\begin{aligned}
N &= \int A'\dot{\bar{\varepsilon}}_m^p \, dt = \int \frac{\left(1-\dfrac{d}{t}\right)N_0}{s_N\sqrt{2\pi}} \, e^{-\frac{1}{2}\left(\frac{\bar{\varepsilon}_m^p - \varepsilon_N}{s_N}\right)^2} \, d\bar{\varepsilon}_m^p = C_2(t-d)\rho\left(\bar{\varepsilon}_m^p\right) \\[2ex]
\rho\left(\bar{\varepsilon}_m^p\right) &= \int \frac{\rho_0}{\sqrt{2\pi}s_N} \, e^{-\frac{1}{2}\left(\frac{\bar{\varepsilon}_m^p - \varepsilon_N}{s_N}\right)^2} \, d\bar{\varepsilon}_m^p \approx \frac{e^{\frac{4}{\sqrt{2\pi}s_N}\left(\bar{\varepsilon}_m^p - \varepsilon_N\right)}}{1 + e^{\frac{4}{\sqrt{2\pi}s_N}\left(\bar{\varepsilon}_m^p - \varepsilon_N\right)}}
\end{aligned}
\right.
$$

$$(4.38)$$

In Eq. (4.38), N_0 represents the increase rate of void number over the thickness direction during the nucleation. $C_2 = N_0 t/\rho_0$ is material constant. Therefore, λ_0 is then denoted as

$$
\lambda_0 = \frac{H}{L} = \frac{C_1 dC_2(t-d)\rho\left(\bar{\varepsilon}_m^p\right)}{t-d} = Cd\rho\left(\bar{\varepsilon}_m^p\right)
$$

$$(4.39)$$

where C is a material constant.

By modifying λ_0 of the Thomason model, the size effect on void coalescence is included. With the increase of grain size, the thickness of the inner layer is decreased and an uneven distribution of voids becomes more significant. Therefore, λ_0 is affected by size effect according to Eq. (4.39) and void coalescence is easier to take place.

4.4.2 FE Simulation and Discussion

As described in Section 4.4.1, the as-observed size effects on the nucleation, growth, and coalescence of voids in damage formation and evolution were incorporated into the extended GTN–Thomason model for sheet metal forming process at meso- and microscale. To verify its applicability, the model was then employed to predict the forming limit and compared with the results obtained in the experimental investigations, as presented in Section 4.2.

4.4.2.1 Parameter Discussion

The established model was first incorporated into ABAQUS/Explicit via the user subroutine of VUMAT. To characterize the failure behavior under different conditions, the coefficients of the established model need to be determined appropriately.

Similar to the analysis in Subsection 4.3.4, some of the coefficients are determined based on the previous research [13,64,65]. Nevertheless, among the parameters to be determined, f_n, C, and $f_f - f_c$ denote the nucleation speed, the difficulty of coalescence, and the acceleration of failure after coalescence, respectively, and they are directly related to failure behavior. These three coefficients are first discussed based on the deformation behavior in the uniaxial tensile test simulation. As an instance, the uniaxial tensile test

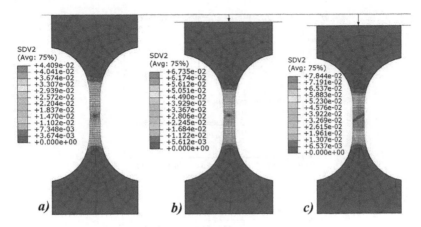

FIGURE 4.36
FE simulation results under different f_n: (a) $f_n = 0.025$; (b) $f_n = 0.05$; and (c) $f_n = 0.1$.

of specimens with a thickness of 0.2 mm and a grain size of 17.4 μm was employed to study the effect of three critical coefficients.

4.4.2.1.1 The Nucleation Coefficient f_n

The simulation results under the conditions of f_n equals 0.025, 0.05, and 0.1, respectively, are shown in Figure 4.36, in which SDV2 stands for the void volume fraction f. The engineering stress–strain curves and the evolution of void volume fraction f and the coalescence indicator χ are shown in Figure 4.37.

It can be observed that the necking and fracture moment starts significantly earlier with the increase of f_n. As illustrated in Figure 4.37b, the increase of f accelerates with the increase of f_n, which directly affects the nucleation of microvoids. As a result, χ also accelerates leading to an earlier initiation of coalescence. In addition, the engineering stress is also identified to have a slight decrease with the increase of f_n from 0.025 to 0.1. But the decrease of flow stress is not significant with a maximum value less than 7% at an engineering strain of 0.2. This is because the increasing porosity softens the flow stress of material. The softening effect is not evident since the void volume fraction is small in the current investigation scenario. As shown in Figure 4.37, f almost doubles when f_n is increased from 0.025 to 0.1, while only a slight decrease of the flow stress is induced.

4.4.2.1.2 The Coalescence Coefficient C

The conditions with $C = 0.25$, 0.5, and 1 were investigated, and the results are shown in Figures 4.38 and 4.39.

Similar to f_n, the increase of C also leads to a decrease of formability. Unlike f_n, C does not affect the increase of void volume fraction f or the flow stress. An evident acceleration of χ can be observed with the increase of C. Therefore, the coalescence starts earlier, leading to a decrease of forming limit.

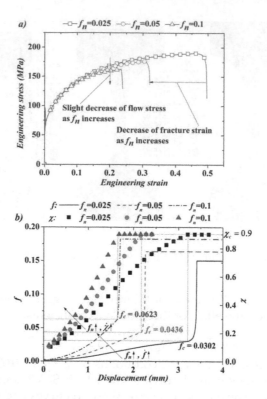

FIGURE 4.37
The effect of f_n on the deformation and damage evolution: (a) Engineering stress–strain curve and (b) Void volume fraction f and coalescence coefficient χ.

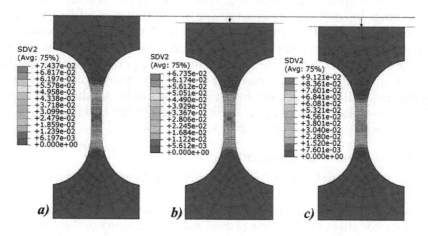

FIGURE 4.38
FE simulation results under different C: (a) $C = 0.25$; (b) $C = 0.5$; and (c) $C = 1$.

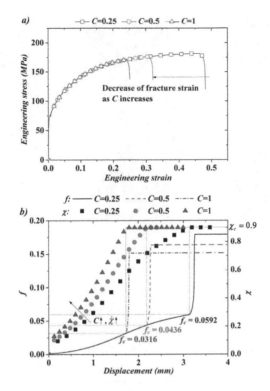

FIGURE 4.39
The effect of C on the deformation and damage evolution: (a) engineering stress–strain curve and (b) void volume fraction f and coalescence coefficient χ.

4.4.2.1.3 The Fracture Acceleration Coefficient $f_f - f_c$

$f_f - f_c$ was identified to be 0.02, 0.12, and 0.3 in FE investigation, and the results are shown in Figures 4.40 and 4.41.

According to Figure 4.41, the coefficient of $f_f - f_c$ only affects the damage evolution after coalescence. With the increase of $f_f - f_c$, the acceleration of void volume fraction f is decreased. Instead of an immediate failure after the coalescence point, the failure progress is gradually delayed with the increase of $f_f - f_c$.

4.4.2.2 Parameter Determination

Obviously, the determined coefficients of f_n, C, and $f_f - f_c$ are linked with the failure behavior of sheet metal closely. They need to be accurately identified before the model can be verified by experiments. Therefore, five-level orthogonal experiments were designed and presented in Table 4.6 to analyze the three factors, i.e., $f_f - f_c$, f_n, and C. In addition, 25 FE simulations for each grain size condition were then performed to obtain the engineering

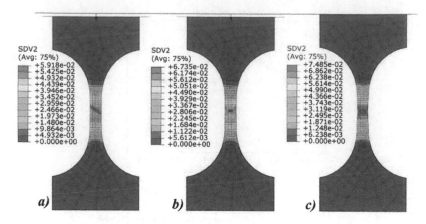

FIGURE 4.40
FE simulation results under different $f_f - f_c$: (a) $f_f - fc = 0.02$; (b) $f_f - fc = 0.12$; and (c) $f_f - f_c = 0.3$.

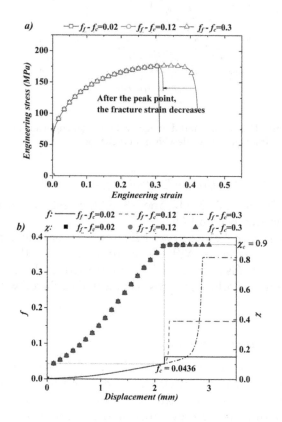

FIGURE 4.41
Effect of $f_f - f_c$ on the deformation and damage evolution: (a) engineering stress–strain curve and (b) void volume fraction f and coalescence coefficient χ.

TABLE 4.6

Design of the Experiments [13]

	f_n	c	$f_f - f_c$
1	0.01	0.1	0.01
2	0.01	0.25	0.06
3	0.01	0.5	0.12
4	0.01	0.75	0.2
5	0.01	1	0.3
6	0.025	0.5	0.3
7	0.025	0.75	0.01
8	0.025	1	0.06
9	0.025	0.1	0.12
10	0.025	0.25	0.2
11	0.05	1	0.2
12	0.05	0.1	0.3
13	0.05	0.25	0.01
14	0.05	0.5	0.06
15	0.05	0.75	0.12
16	0.075	0.25	0.12
17	0.075	0.5	0.2
18	0.075	0.75	0.3
19	0.075	1	0.01
20	0.075	0.1	0.06
21	0.1	0.75	0.06
22	0.1	1	0.12
23	0.1	0.1	0.2
24	0.1	0.25	0.3
25	0.1	0.5	0.01

stress–strain curves. Since the coincidence of the numerically determined and experimental curves is the focus, four reference responses as shown in Figure 4.42 are compared. They are the strain at the maximum stress (R_1), the maximum stress (R_2), the strain at failure (R_3), and the stress at failure (R_4). The square errors $\left(R_i^e, i = 1, 2, 3, 4 \right)$ between the predicted and the experimental results of each grain size condition are added up for each response, and the sums are fitted based on Eq. (4.40) using the least-square method.

$$Y = b_1 + \sum_{i=1}^{3} b_{i+1} X_i + \sum_{i=1}^{3} b_{i+4} X_i^2 + \sum_{i=1}^{2} \sum_{j=i+1}^{3} b_{i+j+5} X_i X_j + \sum_{i=1}^{3} \sum_{j=1, j \neq i}^{3} b_{ij} X_i^2 X_j + b_{17} X_1 X_2 X_3$$

$$(4.40)$$

After the fitted variables of the four reference responses are obtained, the minimum square error was calculated for each response, as shown in

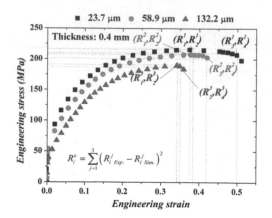

FIGURE 4.42
The schematic of response variables used for parameter determination [13].

Table 4.7. The four points were then averaged with the weights to compute the final results. The fitted parameter values for each thickness were obtained and listed in Table 4.8. The fitted and experimental engineering stress–strain curves are given in Figure 4.43.

4.4.2.3 FLC Construction and Discussion

Upon determination of the parameters, the FE simulations of Holmberg and Marciniak tests were then conducted. As shown in Figure 4.44, the major strain fields obtained by the DIC method are compared with the simulation

TABLE 4.7

The Calculated Parameters Indicating the Minimum Square Error of Each Response Variable [13]

		Weight	f_n	C	$f_f - f_c$
0.1 mm	R_1^e	0.15	0.0722	0.717	0.0752
	R_2^e	0.4	0.0498	0.239	0.193
	R_3^e	0.2	0.0728	0.725	0.0800
	R_4^e	0.25	0.0615	0.387	0.0412
0.2 mm	R_1^e	0.5	0.0568	0.404	0.00426
	R_2^e	0.1	0.110	0.368	0.202
	R_3^e	0.1	0.0725	0.743	0.08
	R_4^e	0.3	0.0165	0.981	0.314
0.4 mm	R_1^e	0.3	0.0342	0.426	0.146
	R_2^e	0.4	0.0366	0.0356	0.0599
	R_3^e	0.1	0.0381	0.366	0.0879
	R_4^e	0.2	0.0234	0.407	0.0830

TABLE 4.8

The Fitted Parameters of the Extended GTN–Thomason Model

Parameters (mm)	f_n	C	$f_f - f_c$
0.1	0.0607	0.445	0.115
0.2	0.0516	0.607	0.125
0.4	0.0334	0.26	0.093

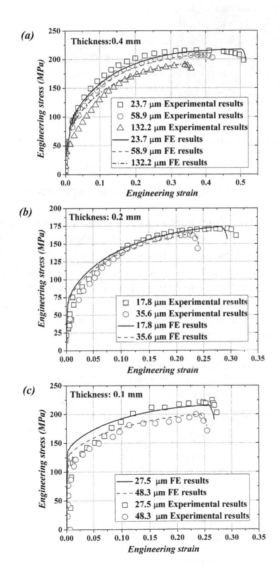

FIGURE 4.43

The fitted engineering stress–strain curves for different conditions [13]: (a) 0.4 mm; (b) 0.2 mm; and (c) 0.1 mm.

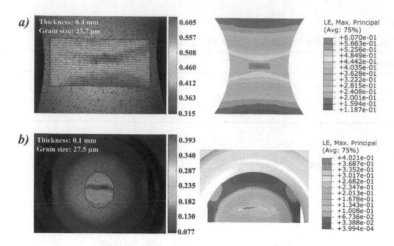

FIGURE 4.44
Strain distribution of the experimental and FE results [13]: (a) Holmberg test and (b) Marciniak test.

results. A similar strain distribution of the experimental and numerical results is also presented.

To analyze the size effect introduced by the extended GTN–Thomason model, the evolution of f and χ is presented in Figure 4.45 for the specimens with a thickness of 0.4 mm and different grain sizes. Before the void coalescence occurs, the increasing rate of f is greater with the smaller grain size. This is because the scale factor is introduced into the nucleation model to describe the decrease of void number at the fracture surface with the decrease of t/d. On the other hand, the increasing rate of coalescence indicator χ increases with the decrease of t/d. This is in accordance with the extended Thomason

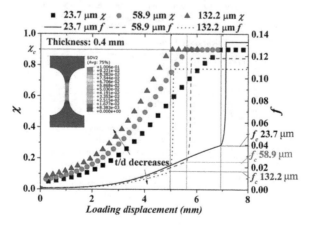

FIGURE 4.45
The size effect on f and χ [13].

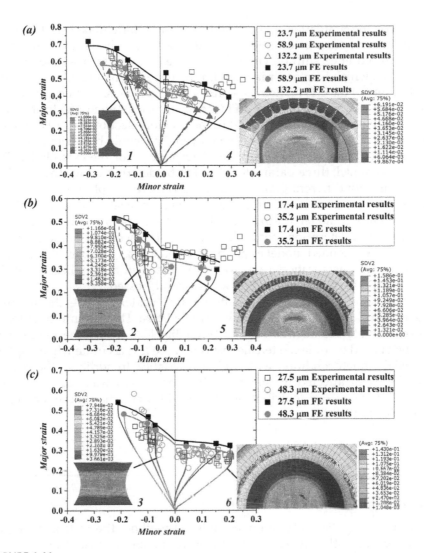

FIGURE 4.46
The prediction of forming limit under different conditions [13]: (a) 0.4 mm; (b) 0.2 mm; and (c) 0.1 mm.

model, considering the fact that the internal layer becomes thinner and thus easier for voids coalescence to start with the decrease of t/d. Since the increasing rate of f is smaller and χ is larger with the decrease of t/d, the critical void volume fraction f_c at the onset of coalescence decreases with t/d, representing that the formability of sheet metal is decreased under meso- and microscale when the size effect becomes more significant.

The major and minor strains are determined as the limit strains when f just reaches f_f and the results are shown in Figure 4.46.

Based on the simulations of both Holmberg and Marciniak tests, the forming limits were then determined and the results are shown in Figure 4.46. The predicted FLCs shift down with the increase of grain size based on the simulation, which is in accordance with the experimental results. Thus, the applicability of the GTN–Thomason model considering size effect is verified. On the other hand, the predicted FLC tends to overestimate the formability compared with the results obtained in the previous section, where the traditional GTN–Thomason model was employed. This could be attributed to the difference of how the parameters are determined. For the extended GTN–Thomason model, three parameters need to be determined for the multiple conditions with different grain sizes. A much more complicated calculation process was involved based on the orthogonal analysis. For the traditional GTN–Thomason model, on the other hand, there are only two parameters that need to be determined for each condition. The accuracy of the obtained results in the extended model might be affected. In fact, the parameter determination remains a critical issue to be addressed in the application of the coupled DF criteria, which need to be further addressed.

Furthermore, the predicted results tend to underestimate the fracture under biaxial stretch condition. It is revealed that Thomason model tends to overestimate the ductility at low triaxiality [67]; hence, the parameters obtained based on uniaxial tensile tests could be less accurate when applied to biaxial stretch condition with higher triaxiality. In addition, the coalescence behavior affected by shear loading component was not considered either. This might also contribute to the error between the experimental and the predicted FLCs.

4.5 Simulation and Validation

To verify the applicability of the established method, the hydroforming of the specimens with a thickness of 0.2 mm was taken as a case study.

By employing a Maximator pump (MHU-M111-G500-2-4500) with adjustable outlet pressure, the hydroforming experimental setup is shown in Figures 4.47 and 4.48. Three sets of dies with increased channel feature dimensions were designed and fabricated. The die was located in the center of die holder plate, and the copper specimen was then placed on the top. Four pieces of apparatus were tightly clamped by eight threaded bolts for reliable sealing. The forming pressure was then provided at a steady speed of 0.2 MPa/s until the occurrence of fracture.

After the experiments, the formed specimens were measured by employing the KEYENCE KS-1100 laser measurement system with a resolution of 0.05 μm. The heights of the hydroformed specimens using Die No. 1 is likely to exceed the measurable scope of this system (–1 to 1 mm); hence, a

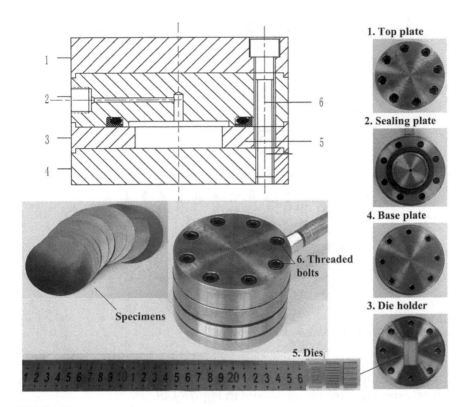

FIGURE 4.47
The hydroforming experimental setup [69].

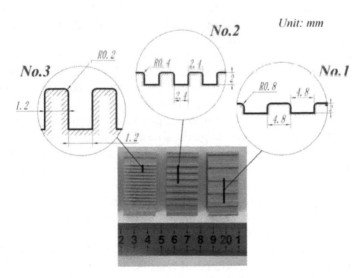

FIGURE 4.48
Geometric dimensions of the hydroforming dies [69].

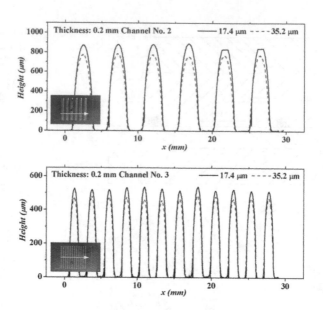

FIGURE 4.49
Measurement of the hydroformed specimens [69].

micrometer caliper with an accuracy of 10 μm was adopted in that case. The laser measurement results are shown in Figure 4.49.

The ultimate height at the onset of failure was identified to decrease with the increase of grain size. The tendency is in accordance with both the experimental and modeling analysis above which the forming limit of sheet metals is affected by size effect and deteriorates when the grain size approaches the geometry dimensions of workpiece.

To characterize the phenomena observed in the hydroforming process, the forming process was simulated based on the extended GTN–Thomason model by using the ABAQUS/Explicit with VUMAT subroutine. The forming process and the failure point are illustrated in Figure 4.50.

The workpiece with a grain size of 0.1 μm was employed to estimate the situation in macroscale when the grain size effect is not significant. An evident reduction of maximum height at the onset of failure with the increase of grain size is observed, which is also consistent with the experiments. By introducing the size effect on void coalescence, the decrease of formability is described using the extended GTN–Thomason model. In addition, the reduction of forming pressure was also simulated by including the size effect on the flow stress of base material.

The numerical results by using the extended GTN–Thomason model were then compared with the experimental results, as shown in Figure 4.51. The predictive results of Swift/Hill criteria, M–K model, and the uncoupled DFCs were also obtained based on the FE simulations without the subroutine.

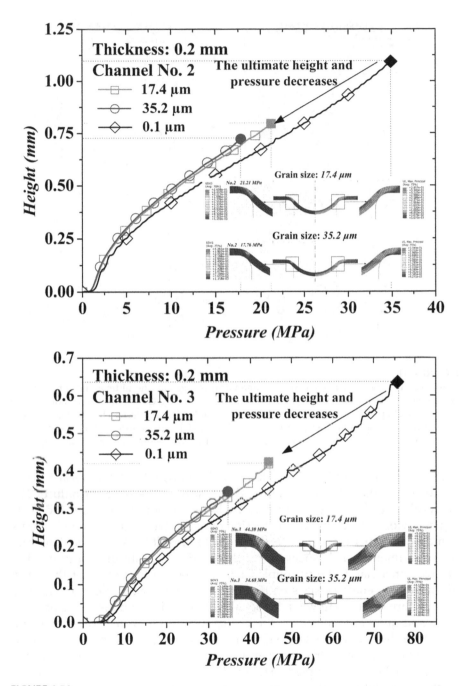

FIGURE 4.50
The bulge height versus pressure curves and the failure points under different grain size conditions in the hydroforming process: (a) Channel No. 2 and (b) Channel No. 3.

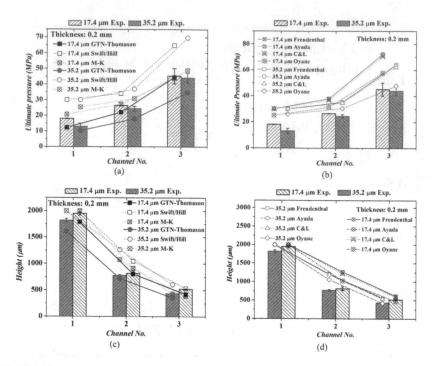

FIGURE 4.51
Comparison of the experimental and the analytical results: (a) The ultimate height predicted according to the Swift/Hill criteria, M–K model and the extended GTN–Thomason model; (b) The ultimate height predicted according to different uncoupled DF criteria; (c) The ultimate pressure predicted according to the Swift/Hill criteria, M–K model and the extended GTN–Thomason model; and (d) The ultimate pressure predicted according to different uncoupled DF criteria.

Obviously, the Swift/Hill model and the uncoupled DF criteria tend to overestimate the ultimate height and pressure at the onset of failure. The classical Swift/Hill criteria could not include the complicated flow and necking behavior induced by the interaction of only several grains in DF process at meso- and microscale. Hence, the predicted result is likely to deviate from the experimental one, especially for the conditions with greater grain size. On the other hand, the uncoupled DF criteria cannot represent the effect of grain and geometry sizes on the necking behavior by solely considering the fracture limit strain when the localized necking happens in the necking zone and quickly followed by the occurrence of DF in hydroforming process. In addition, there is also a significant difference between the predicted FFLC based on these DF criteria and the experimental FFLC. As illustrated in Figure 4.25, the uncoupled DFCs tends to overestimate the FFLC near the plane-strain deformation condition, and this could lead to the overestimation of hydroformed height and pressure at the onset of failure.

To demonstrate the size effect, the same initial surface damage factor f_0^{M-K} value (i.e., the value for the condition with a grain size of 17.4 μm) was used for both conditions using the M–K model. The analytical results with a grain size of 17.4 μm agree with the experiment. The analytical prediction, however, tends to overestimate the ultimate height and pressure with the increase of grain size. Therefore, the M–K model still has room for improvement regarding the handling of the size effect in meso- and microforming. The predictions using the extended GTN–Thomason model are in accordance with the experimental results under different grain size conditions. The reduction of ultimate height and pressure at the onset of failure with the increase of grain size was also described. The extended GTN–Thomason model is able to consider the effect of scale factor on void evolution. Hence, the size effect on forming limit can be well estimated and reasonably explained by using the GTN–Thomason model. Nevertheless, the presented work is only a preliminary research on the failure behavior of ductile materials at meso- and microscale. Many factors, such as stress triaxiality, sheer coalescence behavior, anisotropy, etc., are not considered. The condition with t/d less than two is also not yet discussed in the current work. Therefore, further experimental and theoretical efforts are still needed to address these many eluded and tantalized issues in meso- and microforming.

4.6 Summary

In this chapter, the size effect on the forming limit of sheet metals under different geometry and grain size conditions are presented and discussed. It is well known that size effect affects the deformation behavior of sheet metals at meso- and microscale. There is a lack of an in-depth understanding of size effect on ductile failure behavior in meso- and microforming. In the present research, the well-known FLC was employed as a research objective to analyze the size effect on the failure behavior of sheet metals at multiscales from macro-, meso-, to microscale. The experimental investigations were first performed, and it was found that the FLC shifts down when the grain size approaches the geometry dimensions of the workpiece. In addition, the forming limit results become much more scattered if there are only one or two grains over the thickness direction of sheet metals. The deformation and failure progress was also revealed to be highly related to the geometry and grain sizes. Based on the experimental investigations, various traditional theoretical methods were employed, and their applicability at meso- and microscale are verified. Due to the existence of size effect, some methods such as Swift/Hill model could not obtain satisfying predictive results in its original form. M–K model and the uncoupled DF criteria can describe the FLC and FFLC, respectively, with a reasonable accuracy at different grain size conditions. Nevertheless, their parameters need to be adjusted for each condition. The GTN–Thomason model was found to be suitable for further in-depth discussion as it is able to provide an intuitive description of the nucleation, growth, and coalescence of voids in DF process. Hence, the extended GTN–Thomason model was established and found to describe the decrease of forming limit at meso- and microscales. In addition, the mesoscale hydroforming process was also conducted to verify its applicability. The extended model was shown to characterize the observed geometry and grain size effect mechanism and further to describe the failure behavior.

However, the work presented in this chapter is still a preliminary research on the failure behavior of ductile materials at meso- and microscale. Different from macroscale, the microstructures of materials, including both internal (such as inclusions, grain boundaries, etc.) and surface ones (surface damage, surface roughness, etc.), have an increasingly evident effect on the deformation and failure behavior of materials in meso- and microforming. Further work on the DF at meso- and microscale is critical for promotion of the wide and promising applications of the meso- and microforming of sheet metals in various industrial clusters.

References

1. M. Geiger, M. Kleiner, R. Eckstein, N. Tiesler, U. Engel, Microforming, *CIRP Annals – Manufacturing Technology* 50 (2001) 445–462.
2. F. Vollertsen, Z. Hu, H. Wielage, L. Blaurock, Fracture limits of metal foils in micro forming, in *Proceedings of the 36th International MATADOR Conference*, Springer London, 2010, pp. 49–52.
3. M.W. Fu, W.L. Chan, Geometry and grain size effects on the fracture behavior of sheet metal in micro-scale plastic deformation, *Materials & Design* 32 (2011) 4738–4746.
4. R. Ben Hmida, S. Thibaud, A. Gilbin, F. Richard, Influence of the initial grain size in single point incremental forming process for thin sheets metal and microparts: Experimental investigations, *Materials & Design* 45 (2013) 155–165.
5. T. Furushima, H. Tsunezaki, K.-I. Manabe, S. Alexsandrov, Ductile fracture and free surface roughening behaviors of pure copper foils for micro/meso-scale forming, *International Journal of Machine Tools and Manufacture* 76 (2014) 34–48.
6. J.Q. Ran, M.W. Fu, W.L. Chan, The influence of size effect on the ductile fracture in micro-scaled plastic deformation, *International Journal of Plasticity* 41 (2013) 65–81.
7. J.Q. Ran, M.W. Fu, A hybrid model for analysis of ductile fracture in micro-scaled plastic deformation of multiphase alloys, *International Journal of Plasticity* 61 (2014) 1–16.
8. J. Ran, M. Fu, Applicability of the uncoupled ductile fracture criteria in microscaled plastic deformation, *International Journal of Damage Mechanics*, 25(3) (2015) 289–314.
9. B. Meng, M.W. Fu, C.M. Fu, K.S. Chen, Ductile fracture and deformation behavior in progressive microforming, *Materials & Design* 83 (2015) 14–25.

10. B. Meng, M.W. Fu, C.M. Fu, J.L. Wang, Multivariable analysis of micro shearing process customized for progressive forming of micro-parts, *International Journal of Mechanical Sciences* 93 (2015) 191–203.
11. S.P. Keeler, Determination of forming limits in automotive stampings, *Sheet Metal Industries* 42 (1965) 683–691.
12. G. Goodwin, Application of strain analysis to sheet metal forming problems in the press shop, SAE paper, 680093 (1968).
13. Z.T. Xu, L.F. Peng, M.W. Fu, X.M. Lai, Size effect affected formability of sheet metals in micro/meso scale plastic deformation: Experiment and modeling, *International Journal of Plasticity* 68 (2015) 34–54.
14. M. Geiger, M. Kleiner, R. Eckstein, N. Tiesler, U. Engel, Microforming, *CIRP Annals – Manufacturing Technology* 50 (2001) 445–462.
15. L. Peng, F. Liu, J. Ni, X. Lai, Size effects in thin sheet metal forming and its elastic–plastic constitutive model, *Materials & Design* 28 (2007) 1731–1736.
16. R. Armstrong, I. Codd, R.M. Douthwaite, N.J. Petch, The plastic deformation of polycrystalline aggregates, *Philosophical Magazine* 7 (1962) 45–58.
17. R. Armstrong, The yield and flow stress dependence on polycrystal grain size, in *Yield, Flow and Fracture of Polycrystals*, T.N. Baker, ed., Applied Science Publishers, London, 1983, pp. 1–31.
18. V. Uthaisangsuk, U. Prahl, S. Münstermann, W. Bleck, Experimental and numerical failure criterion for formability prediction in sheet metal forming, *Computational Materials Science* 43 (2008) 43–50.
19. I. Zidane, D. Guines, L. Léotoing, E. Ragneau, Development of an in-plane biaxial test for forming limit curve (FLC) characterization of metallic sheets, *Measurement Science and Technology* 21 (2010) 055701.
20. X. Chen, Z. Yu, B. Hou, S. Li, Z. Lin, A theoretical and experimental study on forming limit diagram for a seamed tube hydroforming, *Journal of Materials Processing Technology* 211 (2011) 2012–2021.
21. U. Engel, Tribology in microforming, *Wear* 260 (2006) 265–273.
22. S. Holmberg, B. Enquist, P. Thilderkvist, Evaluation of sheet metal formability by tensile tests, *Journal of Materials Processing Technology* 145 (2004) 72–83.
23. C. Eberl, R. Thompson, D. Gianola, Digital image correlation and tracking with Matlab, Matlab Central file exchange, 2006.
24. B. Pan, A. Asundi, H. Xie, J. Gao, Digital image correlation using iterative least squares and pointwise least squares for displacement field and strain field measurements, *Optics and Lasers in Engineering* 47 (2009) 865–874.
25. A.G. Atkins, Fracture in forming, *Journal of Materials Processing Technology* 56 (1996) 609–618.
26. M.B. Silva, P.S. Nielsen, N. Bay, P.A.F. Martins, Failure mechanisms in single-point incremental forming of metals, *International Journal of Advanced Manufacturing Technology* 56 (2011) 893–903.
27. G. Centeno, M.B. Silva, L.M. Alves, C. Vallellano, P.A.F. Martins, Towards the characterization of fracture in thin-walled tube forming, *International Journal of Mechanical Sciences* 119 (2016) 12–22.
28. N. Park, H. Huh, S.J. Lim, Y. Lou, Y.S. Kang, M.H. Seo, Facture-based forming limit criteria for anisotropic materials in sheet metal forming, *International Journal of Plasticity* 96 (2017) 1–35.
29. H.W. Swift, Plastic instability under plane stress, *Journal of the Mechanics and Physics of Solids* 1 (1952) 1–18.

30. R. Hill, On discontinuous plastic states, with special reference to localized necking in thin sheets, *Journal of the Mechanics and Physics of Solids* 1 (1952) 19–30.
31. Z. Marciniak, K. Kuczyński, Limit strains in the processes of stretch-forming sheet metal, *International Journal of Mechanical Sciences* 9 (1967) 609–620.
32. A. Freudenthal, *The Inelastic Behavior of Solids*, Wiley, New York, 1950.
33. M. Ayada, T. Higashino, K. Mori, Central bursting in extrusion of inhomogeneous materials, *Advanced Technology of Plasticity* 1 (1987) 553–558.
34. M. Cockcroft, D. Latham, Ductility and the workability of metals, *Journal of the Institute of Metals* 96 (1968) 33–39.
35. T. Wierzbicki, Y. Bao, Y.-W. Lee, Y. Bai, Calibration and evaluation of seven fracture models, *International Journal of Mechanical Sciences* 47 (2005) 719–743.
36. M. Oyane, T. Sato, K. Okimoto, S. Shima, Criteria for ductile fracture and their applications, *Journal of Mechanical Working Technology* 4 (1980) 65–81.
37. A. Gurson, Continuum theory of ductile rupture by void nucleation and growth: Part I-Yield criteria and flow rules for porous ductile media, *Journal of Engineering Materials and Technology* 99 (1977) 2–15.
38. V. Tvergaard, Material failure by void coalescence in localized shear bands, *International Journal of Solids and Structures* 18 (1982) 659–672.
39. V. Tvergaard, A. Needleman, Analysis of the cup-cone fracture in a round tensile bar, *Acta Metallurgica* 32 (1984) 157–169.
40. C. Chu, A. Needleman, Void nucleation effects in biaxially stretched sheets, *Journal of Engineering Materials and Technology (Transactions of the ASME)* 102 (1980) 249–256.
41. J. Besson, Damage of ductile materials deforming under multiple plastic or viscoplastic mechanisms, *International Journal of Plasticity* 25 (2009) 2204–2221.
42. M. Gologanu, J.B. Leblond, J. Devaux, Approximate models for ductile metals containing non-spherical voids—Case of axisymmetric prolate ellipsoidal cavities, *Journal of the Mechanics and Physics of Solids* 41 (1993) 1723–1754.
43. J. Wen, Y. Huang, K.C. Hwang, C. Liu, M. Li, The modified Gurson model accounting for the void size effect, *International Journal of Plasticity* 21 (2005) 381–395.
44. V. Monchiet, G. Bonnet, A Gurson-type model accounting for void size effects, *International Journal of Solids and Structures* 50 (2013) 320–327.
45. A. Weck, D.S. Wilkinson, E. Maire, H. Toda, Visualization by X-ray tomography of void growth and coalescence leading to fracture in model materials, *Acta Materialia* 56 (2008) 2919–2928.
46. A. Hosokawa, D.S. Wilkinson, J. Kang, E. Maire, Onset of void coalescence in uniaxial tension studied by continuous X-ray tomography, *Acta Materialia* 61 (2013) 1021–1036.
47. W.H. Liu, Z.T. He, J.G. Tang, Z.J. Hu, D.T. Cui, The effects of load condition on void coalescence in FCC single crystals, *Computational Materials Science* 60 (2012) 66–74.
48. Y. Zhang, Z. Chen, On the effect of stress triaxiality on void coalescence, *International Journal of Fracture* 143 (2007) 105–112.
49. R.E. Rudd, J.F. Belak, Void nucleation and associated plasticity in dynamic fracture of polycrystalline copper: An atomistic simulation, *Computational Materials Science* 24 (2002) 148–153.
50. G.P. Potirniche, M.F. Horstemeyer, G.J. Wagner, P.M. Gullett, A molecular dynamics study of void growth and coalescence in single crystal nickel, *International Journal of Plasticity* 22 (2006) 257–278.

51. P.F. Thomason, A three-dimensional model for ductile fracture by the growth and coalescence of microvoids, *Acta Metallurgica* 33 (1985) 1087–1095.
52. P.F. Thomason, Three-dimensional models for the plastic limit-loads at incipient failure of the intervoid matrix in ductile porous solids, *Acta Metallurgica* 33 (1985) 1079–1085.
53. P. Thomason, Ductile fracture of metals, Pergamon Press plc, Ductile Fracture of Metals(UK), 1990, p. 219.
54. T. Pardoen, J.W. Hutchinson, An extended model for void growth and coalescence, *Journal of the Mechanics and Physics of Solids* 48 (2000) 2467–2512.
55. F. Scheyvaerts, T. Pardoen, P. Onck, A new model for void coalescence by internal necking, *International Journal of Damage Mechanics* 19 (2010) 95–126.
56. D. Fabrègue, T. Pardoen, A constitutive model for elastoplastic solids containing primary and secondary voids, *Journal of the Mechanics and Physics of Solids* 56 (2008) 719–741.
57. Z. Zhang, C. Thaulow, J. Ødegård, A complete Gurson model approach for ductile fracture, *Engineering Fracture Mechanics* 67 (2000) 155–168.
58. C.-K. Oh, Y.-J. Kim, J.-H. Baek, Y.-P. Kim, W. Kim, A phenomenological model of ductile fracture for API X65 steel, *International Journal of Mechanical Sciences* 49 (2007) 1399–1412.
59. Y. Yan, Q. Sun, J. Chen, H. Pan, The initiation and propagation of edge cracks of silicon steel during tandem cold rolling process based on the Gurson–Tvergaard–Needleman damage model, *Journal of Materials Processing Technology* 213 (2013) 598–605.
60. H. Li, M. Fu, J. Lu, H. Yang, Ductile fracture: Experiments and computations, *International Journal of Plasticity* 27 (2011) 147–180.
61. M. Abbasi, B. Bagheri, M. Ketabchi, D.F. Haghshenas, Application of response surface methodology to drive GTN model parameters and determine the FLD of tailor welded blank, *Computational Materials Science* 53 (2012) 368–376.
62. V. Uthaisangsuk, U. Prahl, W. Bleck, Characterisation of formability behaviour of multiphase steels by micromechanical modelling, *International Journal of Fracture* 157 (2009) 55–69.
63. D. Simulia, ABAQUS 6.11 analysis user's manual, Abaqus 6.11 Documentation, 2011, 22.22.
64. R. Chhibber, P. Biswas, N. Arora, S. Gupta, B. Dutta, Micromechanical modelling of weldments using GTN model, *International Journal of Fracture* 167 (2011) 71–82.
65. Z. Chen, X. Dong, The GTN damage model based on Hill'48 anisotropic yield criterion and its application in sheet metal forming, *Computational Materials Science* 44 (2009) 1013–1021.
66. A.A. Benzerga, J. Besson, A. Pineau, Coalescence-controlled anisotropic ductile fracture, *Journal of Engineering Materials and Technology* 121 (1999) 221–229.
67. A.A. Benzerga, Micromechanics of coalescence in ductile fracture, *Journal of the Mechanics and Physics of Solids* 50 (2002) 1331–1362.
68. Z.T. Xu, L.F. Peng, X.M. Lai, M.W. Fu, Geometry and grain size effects on the forming limit of sheet metals in micro-scaled plastic deformation, *Materials Science and Engineering A* 611 (2014) 345–353.
69. L.F. Peng, Z.T. Xu, M.W. Fu, X.M. Lai, Forming limit of sheet metals in meso-scale plastic forming by using different failure criteria, *International Journal of Mechanical Sciences* 120 (2017) 190–203.

5

Meso- and Microforming of Sheet Metals for Making Sheet Metal Parts and Components

5.1 Introduction

Sheet metal forming has been widely used in the manufacture of macroproducts for various applications in different industrial clusters such as automotive, aerospace, consumer electronics, biomedical, etc., for its low cost, high productivity, less energy consumption, good formability of materials, and further good feasibility for a large-scale mass production of microparts. With the overwhelming trend of product miniaturization in electronics and other disciplines, the growing demand for microparts is becoming critical. How to quickly and efficiently produce a large-scale production of microparts to meet the crucial demand from industries has become a nontrivial issue to be addressed. Meso- and microscaled sheet forming provides an efficient solution. With the scaling down of geometry size of sheet metal-formed parts, there is a variation of deformation behaviors and mechanical behaviors of sheet metals, which exhibit sizes dependent and different from those in macroscale. Miniaturization affects not only the deformation mechanisms inside the microparts during the process but also the interfacing mechanism on the tooling–workpiece interaction. As a result, the deformation behaviors in microforming of sheet metals would be different from those in macroforming. On the other hand, microforming thus faces challenges, including difficulties in achieving desirable geometries, accurate dimensional accuracy and tolerance, and tailor-needed product properties and quality due to many unknowns, uncertainty, and the scatter of behaviors, phenomena, results, etc.

Till now, there are a lot of investigations on microforming sheet metals, and the results provide good guidance in mass production of microsheet metal parts in industries. Among all the microsheet forming processes, micro bending, micro deep drawing, microsheet hydroforming, and microflexible sheet forming by soft punch are the most concerned processes. From the process aspect, this chapter presents the state of the art of the four microsheet forming processes and introduces the characteristics of each process and forming parameters.

5.2 Microsheet Bending Process

In microsheet bending process, the metal sheet mainly experiences bending deformation in the corner for formation of microfeatures. The deformation springback is more likely to occur at the corners of microfeatures, which remarkably impacts the forming quality of sheet metal. To address this issue, various research projects have been conducted by concentrating on the studies of microbending behaviors of sheet metals. Among the pioneering studies, Suzuki et al. [1] observed that the nondimensional bending moment increases with the thickness decrease of pure aluminum foil from 51 to 24 μm in microbending experiments. In addition, Liu et al. [2] explored the effects of material thickness and microstructural grain size on the springback angle in the three-point bending experiment of pure copper sheet foils. The springback angle is identified to increase with the reduction of thickness. Later, Wang et al. [3] performed U-bending tests to investigate the influences of thickness, grain size, and punch radius on the springback angle and developed a finite element (FE) model based on the surface layer model to predict the springback angle. All of these efforts advance the understanding of the deformation in microbending processes.

5.2.1 Loading and Displacement in Microbending Process

Figure 5.1 shows a setup for microbending experiment and the detailed punch and die set in the setup of microbending experiment. Figure 5.1a presents a monitoring system with a digital camera. A camera with high resolution is placed in front of the bending tool and is used to record the deformation behavior during the bending process continuously. Figure 5.1b illustrates the V-bending experimental setup from Xu et al. [4]. The key design parameters of the tool include punch angle, die angle, punch fillet, die fillet, die shoulder fillet, and die depth. In Xu's research, three punch angles, i.e., 30°, 60° and 90°, were selected to test the specimens with different thicknesses and grain sizes.

By using the V-bending setup, Xu et al. [4] conducted the bending of pure copper sheets with a thickness of 0.1, 0.2, and 0.4 mm using the punches with different punch angles, as shown in Figure 5.2. In the microbending process, after placing the specimen on the die carefully, the punch was moved down manually until it touched the surface of the specimen. The punch then moved at a specified velocity and the bending experiment began. After the punch reached the preset stroke, the punch returned to its original place at the same speed. During the loading and unloading process, a digital camera was employed to take continuous images.

The loading–displacement curves were recorded during the microbending tests and are shown in Figure 5.3. The bending process can be divided into three stages based on the loading–displacement curves. The first one is the

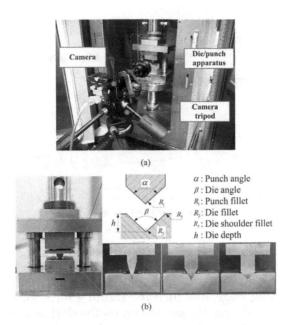

FIGURE 5.1
(a) Setup of microbending experiment and (b) the punch and die set in the microbending experiment [4].

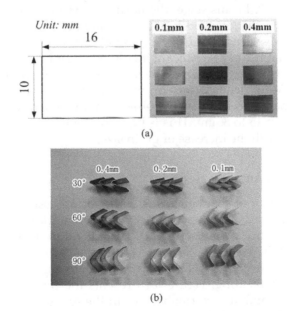

FIGURE 5.2
(a) Geometric dimensions of the bending specimens and (b) the specimens after the microbending test [4].

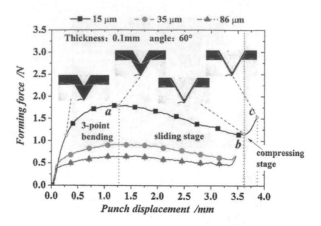

FIGURE 5.3
Force–displacement curves in microbending process [4].

three-point bending stage, in which the forming force increases gradually as the punch moves down. The second one is the sliding stage (from point *a* to *b*, as illustrated in Figure 5.3). As soon as the bending angle of the specimen reaches a critical value, the specimen gradually slides into the groove. The contacting area between the specimen and the groove inner surfaces increases. The third stage is the compressing stage, from point *b* to *c*, as illustrated in Figure 5.3. In this stage, the normal distance between punch and die approaches the thickness of sheet metal. The specimen is pressed against the walls of groove, leading to an increase of forming force. The test stops before the appearance of a significant increase of forming force, at which the compressive plastic deformation instead of bending of specimen becomes a major responding deformation. In addition, it is also observed that the load–displacement curve of the copper sheet shifts down with the increase of grain size from 15 to 86 μm. This is because the flow stress of the copper sheet decreases with the increase of grain size.

5.2.2 Springback Angle Analysis

Springback behavior of sheet metals is an important issue that needs to be extensively addressed, as it directly affects the geometrical and dimensional accuracy of the fabricated sheet metal parts. Bending process is also one of the most widely used approaches to study the springback of sheet metals. Taking the pure copper sheet as an example to analyze the springback behavior in microbending, Figure 5.4 presents the digital images of copper sheets before and after springback from the research done by Xu et al. [4]. The bended angles, before and after bending springback, were measured from the recorded pictures, by an edge detection algorithm, and the springback angle was then calculated. The specimen thickness *t*, grain size

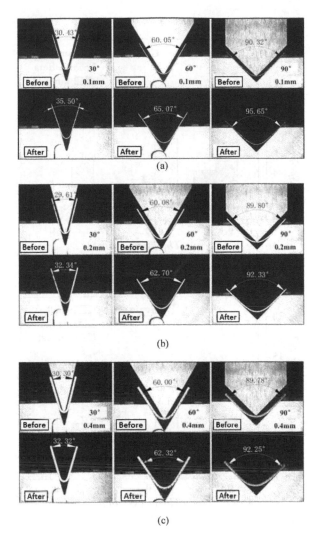

FIGURE 5.4
Digital images of pure copper specimens before and after springback [4].

d, deformation angle, and the ratio of thickness to grain size t/d affect the springback angle of the copper sheets obviously. In the following, the influences of these parameters are discussed in detail.

5.2.2.1 Effect of Specimen Thickness

Figure 5.5 illustrates the effect of thickness on the springback angle. It can be observed that the springback angle in the bending test decreases with the increase of thickness for the specimens with a similar grain size. It is well known that the sheet metal under bending can be divided into outer and

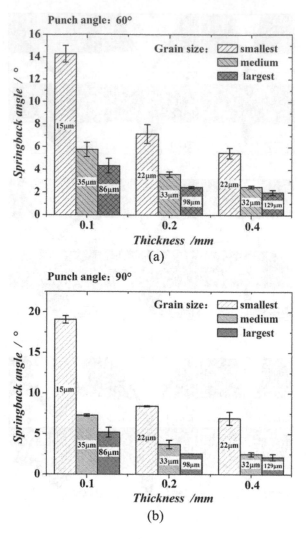

FIGURE 5.5
The effect of thickness on the springback angle of the specimens bent with different punch angles [4].

inner layers by the neutral layer. The outer and inner layers are subjected to tensile and compression stresses, respectively. For the sheet metals bended to the same angle, the maximum tensile and compression strain in the deformation zone increase with the thickness or the distances from the outer and inner surfaces to the neutral layer. As a result, the elastic proportion to the overall deformation decreases with the increase of thickness, which leads to smaller elastic recovery, i.e., the springback angle, considering that the specimens are bended to the same angle.

5.2.2.2 Effect of Grain Size

The influence of grain size on the springback of the specimens under different conditions is shown in Figure 5.6. It is observed that the springback angle decreases with the increase of grain size in each condition. This is because the yield strength and flow stress have been shown to decrease with the increase of grain size. As the specimen is subjected to the same punch angle, the deformation is similar for the specimens with different grain sizes. On the one hand, the specimens with greater grain sizes have lower yield strength. Therefore, they are easier to have plastic deformation. The elastic recovery thus becomes less obvious with the increase of grain size. On the other hand, the decreasing flow stress also results in the smaller elastic recovery in the plastic bending area during the unloading process. The springback angle thus decreases with the increase of grain size.

5.2.2.3 Effect of Deformation Angle

Figure 5.7 illustrates the springback results of the specimens bent by punch with different punch angles. The springback angle of the specimen with the same thickness and grain size is shown to increase with the increase of deformation angle. This is because the bending deformation is less severe with the increase of deformation angle. As a result, the intensity of plastic deformation decreases with the increase of deformation angle, leading to the increasing proportion of elastic strain. The recoverable deformation thus takes a greater proportion in the overall deformation. Hence, the elastic recovery is more significant when the bending load is released, leading to an increase of springback angle.

5.2.2.4 Effect of t/d

The ratio of thickness to grain size, t/d, can be used to reflect the number of grains over the thickness direction of the specimen. With the decrease of t/d, the surface grains take a greater proportion, leading to a more evident size effect in meso- and microscale. Therefore, t/d is considered as one of the most important parameters to characterize the size effect of sheet metal in meso- and microforming. The influence of t/d on the springback of specimens is shown in Figure 5.8.

The experimental results in Figure 5.8 show a significant combining effect of both the geometric dimension and grain size. It can be clearly observed that the springback angle increases with t/d for the specimens with the same thickness. On the other hand, the springback behavior of the specimens with greater thickness is also found to be less obvious. The t/d ratio can be used to characterize both the dimensional and grain size effects of the specimen itself. As shown earlier, based on the surface layer model, the yield strength and flow stress of specimens decrease with t/d due to the increasing proportion

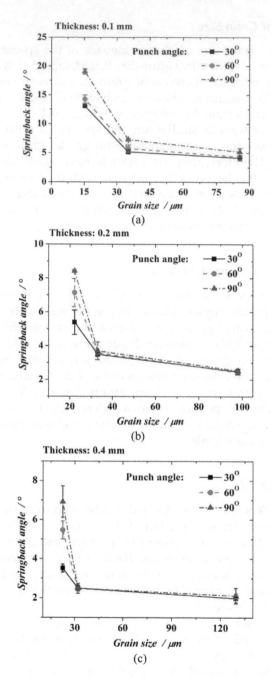

FIGURE 5.6
The effect of grain size on the springback angle of specimens with different thicknesses [4].

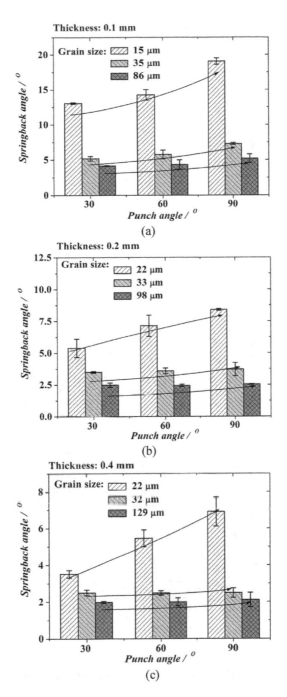

FIGURE 5.7
The effect of punch angle on the springback angle of the specimens with different thicknesses [4].

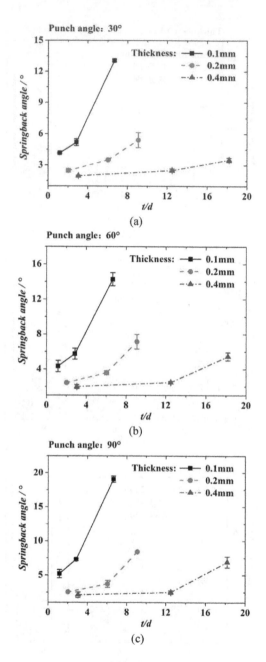

FIGURE 5.8
The effect of t/d on the springback angle of the specimens bent with different punch angles [4].

of surface grains. As a result, the springback behavior becomes less obvious since the elastic-recoverable deformation takes a lesser proportion. Therefore, t/d can illustrate how the geometric and grain sizes of specimen affect the springback behavior during the bending process. In addition, the dimensional factors of the die set also have an obvious influence on the springback behavior. As discussed earlier, the interactive effect of die, punch, and specimen dimensions is also an important influencing factor in the microbending process. Bending area becomes more severe with the increase of sheet metal thickness and the reduction of punch angle. Consequently, the deformation intensity in the bending area increases, leading to the less proportion of elastic-recoverable deformation and the decrease of springback angle, as shown in Figure 5.8.

5.3 Micro Deep Drawing Process

In deep drawing process, a flat sheet metal plate with an appreciate size is pressed into the die cavity to form a cup-shaped part. In micro deep drawing, more new challenges significantly emerge due to the size effect compared with macro deep drawing. Size effect causes an increase of friction and thus hinders the material flow. Therefore, the formability in micro deep drawing is not so good compared with that in macroforming process. To explore the size effect affected micro deep drawing process, many efforts have been provided to delve this unique process. Among them, Saotome et al. [5] conducted the micro deep drawing experimentally to produce the drawn cups with the punch diameter from 1.0 to 10.0 mm by using (steel plate cold elongation) SPCE foils with the thickness of 0.1 mm. They clarified that the forming limit decreases with the increasing ratio between the relative punch diameter to thickness. Vollertsen [6] carried out micro deep drawing experiments using a pure aluminum sheet with a thickness of 0.02 mm and showed that the lubrication was improved by increasing the punch speed, leading to an improvement of the fracture limit and expansion of the forming range. To improve formability, die coating (Shimizu et al. [7]), resistance heating (Tanabe and Yang [8]), and redrawing (Manabe et al. [9]) have been carried out. Irthiea et al. [10] developed a micro deep drawing with flexible dies using rubber material in the die and succeeded in forming cups with an aspect ratio of 1.4 using SUS304 foils, with the thickness from 0.06 to 0.15 mm. In addition, Vollertsen et al. [11] developed a micro deep drawing process with a pulsed laser, and succeeded in clarifying the mechanism behind the process and forming of the microcups using pure aluminum, copper alloy, and stainless-steel foils with a thickness of 0.02 and 0.05 mm. Furthermore, Huang et al. [12] proposed an ultrasonic-assisted drawing system and found that the limit drawing ratio of the deep drawing process can be increased significantly aided by ultrasonic vibration.

5.3.1 Deformation Behavior

Deformation load is an important parameter in micro deep drawing. The relationship between deformation load and punch stroke in each stage of multistage micro deep drawing is shown in Figure 5.9. In the multistage micro deep drawing conducted by Li et al. [13], the diameter and thickness of the original copper blank are 3 and 0.2 mm, respectively. The sheet blank was drawn to the cylindrical die with the diameter of 1.7 mm by the first domed punch with a diameter of 1.3 mm to obtain the semifinished part. The first drawn part was then redrawn to the final micropart with the inner diameter of 1.0 mm and a height of 2.07 mm.

In Figure 5.9a, the deformation load increases first to the maximum point and then decreases to form the first peak in the first micro deep drawing operation. When the thick flange is drawn into the clearance between punch and die, the second peak occurs. However, there are not obviously two peaks in the load–stroke curves as shown in Figure 5.9b. In addition, the peak loads with different annealing temperatures in the first micro deep drawing are

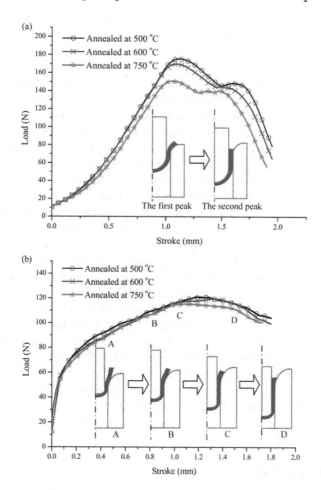

FIGURE 5.9
Load–stroke curve: (a) the first operation and (b) the second operation [13].

much larger than those in the second stage. The main reason is that the deformation is decreased in the second stage. Furthermore, it can be observed that the deformation load decreases with the increase of annealing temperature, while the decrease of deformation load in Figure 5.9b is considerably slighter than that in Figure 5.9a.

The difference of deformation loads at different annealing temperatures can be explained by the influence of surface grain size and the grain boundary strengthening behavior. Figure 5.10 shows the microstructures of the final parts in the second-stage micro deep drawing, using the blank annealed at 500°C and 750°C, respectively. From the figure, it can be seen that the corner of the part annealed at 750°C consists of only one or two grains along the thickness direction. Therefore, the blank with a larger grain size and less grains across its thickness has a relatively low density of grain boundaries, which results in a lower restriction to deformation, and low deformation load. However, it can be seen that the grain size has a less significant effect on the deformation load during the second micro deep drawing stage.

In addition, the material anisotropy, namely the different material properties induced by the deformation texture in different directions, exists. When the earing appears in the first stage, the ear with a larger difference in height was developed after the second-stage drawing, as shown in Figure 5.11a. It is mainly due to the greater material deformation in multistage drawing than that in single-stage drawing. When the earing and misalignment exist simultaneously and make the height difference of semifinished product more considerable in the first-stage micro deep drawing, the final part with a larger

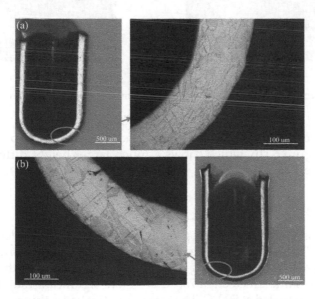

FIGURE 5.10
Microstructures of the final parts in the second micro deep drawing operation. Annealed at (a) 500°C and (b) 750°C [13].

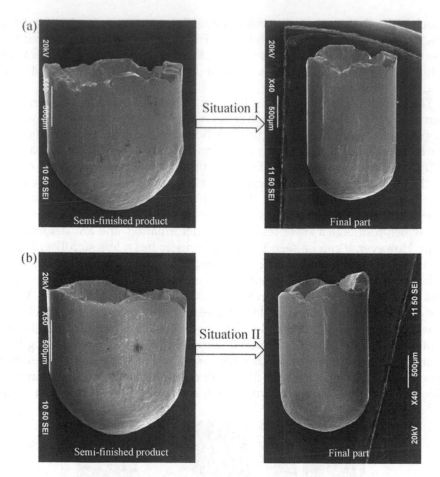

FIGURE 5.11
Scanning electron microscope images of microparts in each stage of the micro deep drawing process: (a) situation I and (b) situation II [13].

height difference was made after the second-stage micro deep drawing, as shown in Figure 5.11b, which shows a much larger height difference than that in Figure 5.11a.

5.3.2 Thickness Variation

The thickness variation of microparts in each stage of micro deep drawing is shown in Figure 5.12. It can be observed that the nonuniform thickness distribution and severe thinning around the punch corner of the micropart made by the blank annealed at 750°C are more significant than that drawn by the blank annealed at 500°C. The maximum thickness thinning is approximately 25% and 35% of the initial blank thickness for the microparts drawn by the blank annealed at 750°C in the first- and second-stage micro deep

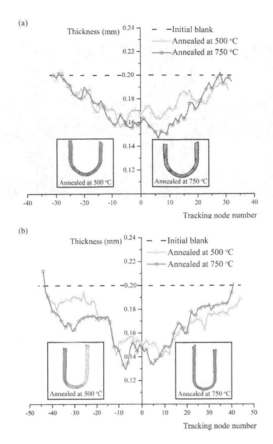

FIGURE 5.12
Thickness variation of microparts: (a) the first stage and (b) the second stage [13].

drawing, respectively, larger than the values of about 20% and 30% for the microparts drawn by using the blank annealed at 500°C.

The nonuniform thickness of microparts with different annealing temperatures is mainly caused by the surface roughening induced by plastic deformation in multistage micro deep drawing. From Figure 5.13, it can be clearly observed that the surface roughness at the bottom part in the second stage is significantly larger than that in the first stage with the increase of the annealing temperature from 500°C, 600°C to 750°C. Considering the larger surface roughness of the micropart drawn by the blank annealed at 750°C, it is affected by grain size, grain orientation, and grain rotation during plastic deformation. The blank annealed at 750°C has a few grains or single grain along the thickness direction and less constrained in deformation due to a large fraction of surface grains. These grains can rotate out of their original position by gliding along the grain boundaries to result in a rougher surface. In addition, from the research of Li et al. [13], it is also demonstrated that the surface roughness

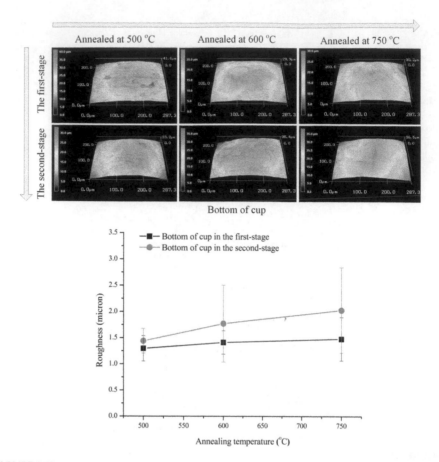

FIGURE 5.13
Surface roughness at the bottom of the cup drawn using different annealed blanks in the two continuous stages of micro deep drawing [13].

at the wall of the cup drawn by different annealed blanks is almost the same in the first-stage drawing, while the surface roughness at the wall of the cup in the second stage is lower than that in the first stage. The decreased unilateral clearance in the second stage has an ironing effect on the wall of the cup, which makes its surface roughness smaller than that in the first stage.

5.4 Microsheet Metal Hydroforming

Generally, metal hydroforming can be classified into tube hydroforming and sheet metal hydroforming. Tube hydroforming has a good potential for making various parts for different industrial clusters, and thus the process has been extensively studied theoretically, numerically, and experimentally [14,15]. Nevertheless, the development of sheet metal hydroforming is much slower since the process is very complex and needs the equipment with a great capacity and high investment [16]. However, in recent years, a lot of impressive achievements have been made in sheet metal hydroforming. To name a few, Hein et al. [17–19] proposed a combined process of hydrofroming and welding of metal pairs; Ahmetoglu et al. [20] developed a forming method using viscous material instead of water in hydroforming process. Their work demonstrated that sheet metal hydroforming is a promising approach to fabricate sheet metal parts with complex shape. In addition, Joo et al. [21] investigated the feasibility of producing microscale structures by forming ultrathin metal foils using isostatic pressing. Mahabunphachai et al. [22] studied the fabrication of microchannels using the internal fluid pressure hydroforming. When the feature size is decreased into meso- and microscale, the meso- and micro hydroforming have advantages in die assembly design and making, as it does not have problems such as positioning and alignment issues and only one rigid die is required to be manufactured. Therefore, the process can save time to market and production cost [23]. The advantages and the promising potential of microsheet hydroforming have been fully demonstrated in many previous literatures.

5.4.1 Processing Procedure

A hydroforming experimental setup developed by Xu et al. [24] is shown in Figure 5.14a, where the fluid serves as a punch. For the die, ten channels numbered from 1 to 10 were designed with different widths from 0.8 to 1.6 mm. The radius of the fillet has the size from 0.2 to 0.6 mm, and the height has a dimensional range from 0.4 to 1.2 mm. The oblique angles are 0° and 15°. Figure 5.14b shows these details. The die is placed in the center of the die holder. The sheet specimen with a diameter of 70 mm is located between the oil chamber and the die. The roof board is placed on the top, and all the pieces are tightly clamped together by eight threaded bolts. At each interface, the high strength rubber seal rings, which could bear a static pressure up to 200 MPa, are used to prevent oil leakage. Using a super high pressure oil pump, the hydroforming experiments were conducted under different pressures, and thus workpieces with microchannels were made for SS304 sheet metal with a thickness of 0.1 mm. Figure 5.15 shows hydroformed workpieces under different pressures. Based on the experimental results, fracture first occurred at Channels 5 and 10 for each die if the pressure was raised to 132 and 166 MPa, respectively.

Figure 5.16 presents the 2D profiles of hydroformed SS304 sheets with a thickness of 0.1 mm. The workpiece was measured along three different scan paths, and the height of each channel was averaged. It was found that the height variation along the channel's length direction is not significant.

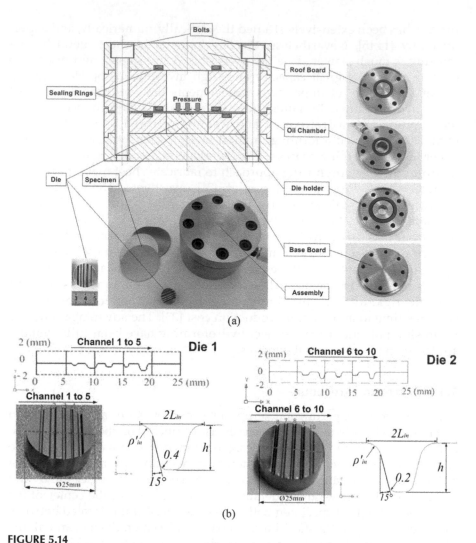

FIGURE 5.14

(a) Experimental setup of microsheet metal hydroforming and (b) the picture of the dies with microchannel arrays (five channels in each) [24].

In Figure 5.16a, the peak of the profiles of Channel 2 at a pressure of 95 MPa or even higher is flat, which means that the workpiece is compressed to the bottom when the pressure is larger than 95 MPa. The difference between the dome height of the workpiece and the die at Channel 2 may be caused by the springback of sheet metal. In Figure 5.16b, it is observed that the right side of Channel 10 is 0.2 mm lower than the rest of the parts. As a result, the height between the dome point and the lower side was measured for Channel 10 to compare with the analytical result, although the forming situation of Channel 10 is different from the model established earlier.

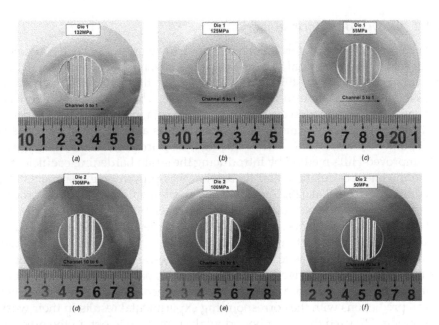

FIGURE 5.15
The hydroformed workpieces [24].

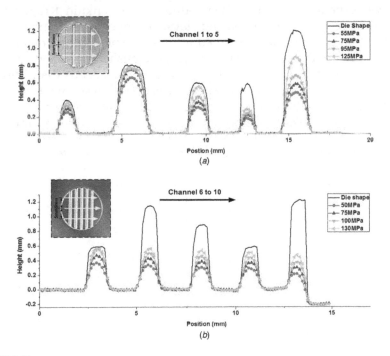

FIGURE 5.16
Measurement of workpieces under different hydropressures [24].

5.4.2 Analysis of Deformation

In terms of analysis of deformation process, many works have been conducted for development of the analytical models for analyzing the sheet metal hydroforming process. In the 1950s, Hill [25] studied the analytical methods to describe the deformation in hydraulic bulge test. The explicit formulae were obtained for determination of stresses in a metal diaphragm bulged plastically by lateral pressure. In terms of calculation, it was assumed that the locus of each point on the sheet was a circle during the deformation. Chakrabarty et al. [26] improved Hill's method by introducing the strain hardening coefficient in dome thickness calculation, and their work was proven by experiments [27]. In addition, Shang et al. [28,29] studied the sheet metal hydroforming process by establishing mathematical models for hydraulic bulge tests with and without draw-in allowed. A good correlation between theoretical and experimental results was observed. Furthermore, Kruglov et al. [30] developed a mathematical model for the superplastic free forming of edge-welded titanium envelopes. Kruglov's model was derived based on the principal equations of the membrane theory, and a good agreement was obtained by comparing the theoretical predictions with the corresponding experimental results in their work. Assempour [31] used the upper bound analysis method to obtain the pressure equation with kinematically admissible velocity field. Koc et al. [32] summarized different approaches for analysis of hydraulic bulge tests and compared them with experimental results. They found that the Panknin–Kruglov model had the best accuracy at both cold and elevated temperature conditions.

In addition to the earlier described efforts in exploration of deformation and modeling of the process, Xu et al. [24] established an analytical model considering the influence of fillet and the inhomogeneous distribution of stress and strain through the thickness direction of sheet metals to predict the height of the hydroformed microchannels in plane-strain condition. As shown in Figure 5.17, the deformed workpiece can be divided into three

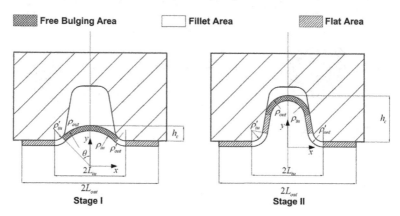

FIGURE 5.17
Two stages of the hydroforming process [24].

areas according to the loading condition in the forming process. They are (a) free bulging area, in which the material is freely bulged into the channel; (b) fillet area, where the material contacts the fillet of the channel; and (c) the flat area, where the material in this area is compressed to flat zones of the channel. The deformation process can be assumed to contain two stages. In Stage I, the angle of the free bulging area θ will increase with the hydraulic pressure, as shown in Figure 5.18. If θ is increased to $\pi/2-\alpha$, the deformation of Stage II will occur. In Stage I, the sheet begins to flow into the channel and contacts the fillet; while in Stage II, the sheet is bulged into the channel and contacts the inner wall of the channel, as illustrated in Figure 5.19.

According to the principle of volume constancy and the assumption of plane-strain condition, the equivalent strain and stress can be obtained, respectively, from Mises yield criterion as follows:

$$\bar{\varepsilon} = \frac{2}{\sqrt{3}}\sqrt{\left(\varepsilon_\varphi\right)^2} = \frac{2}{\sqrt{3}}\left|\varepsilon_\varphi\right| \tag{5.1}$$

$$\bar{\sigma} = \frac{\sqrt{3}}{2}\left|\sigma_\varphi - \sigma_\rho\right| \tag{5.2}$$

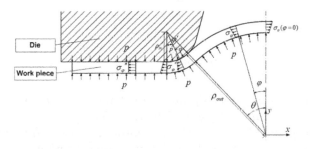

FIGURE 5.18
Force and deformation in Stage I [24].

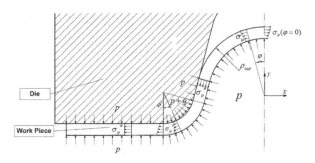

FIGURE 5.19
Force's boundary condition in Stage II [24].

Substituting Eqs. (5.1) and (5.2) into Ludwigson equation, the following is obtained:

$$\frac{\sqrt{3}}{2}\left|\sigma_\varphi - \sigma_\rho\right| = K_1 \left(\frac{2}{\sqrt{3}}\left|\varepsilon_\varphi\right|\right)^{n_1} + e^{K_2 + n_2 \frac{2}{\sqrt{3}}\left|\varepsilon_\varphi\right|} \tag{5.3}$$

Considering Eq. (5.3), the main procedure of the modeling approach is

a. Establishing the relationship between stress and pressure according to the force balance analysis.
b. Figuring out the correlation between strain and the characteristic geometric variables based on geometry analysis.
c. Developing the relation between the hydraulic pressure p and the geometric parameters (e.g. the dome height h_c) by substituting the results obtained earlier into Eq. (5.3).

In Stage I, $\theta < \pi/2 - \alpha$, according to the assumptions, the height can be described as follows:

$$h_c = \frac{L_{in}}{\sin\theta}(1 - \cos\theta) \tag{5.4}$$

where

$$\theta = \sin^{-1}\left(\frac{L_{in}}{\rho_{out} + \rho'_{in}}\right) \tag{5.5}$$

Supposing that the relationship between P and ρ_{out} is obtained, the dome height h_c can be calculated by Eqs. (5.4) and (5.5).

Subsequently, the stress and strain conditions in the free bulging area, the fillet area, the flat area, and the deformation condition of workpiece can be studied to establish the relationship between P and ρ_{out} according to the approach proposed and described earlier. Due to the symmetric condition, half of the workpiece is analyzed and discussed.

In Stage II, the following relations of the geometric parameters can be derived.

$$\theta = \frac{\pi}{2} - \alpha \tag{5.6}$$

$$\rho_{out} \leq \frac{L_{in}}{\sin\left(\frac{\pi}{2} - \alpha\right)} - \rho'_{in} \tag{5.7}$$

$$h_c = \left(L_{in} - \rho_{out}\sin\theta - \rho'_{in}\sin\theta\right)\tan\theta + \left(\rho_{out} + \rho'_{in}\right)(1 - \cos\theta) \tag{5.8}$$

θ is equal to $\pi/2 - \alpha$ during the deformation process in Stage II. With the material flowing into the channel, ρ_{out} is getting smaller and smaller. The height of the workpiece h_c can be represented by the function ρ_{out} in Eq. (5.8) due to the geometric constraint.

By inputting the length of the workpiece L_{out}, the width L_{in}, the oblique angle α, and the fillet radius ρ'_{in} of the microchannels into the analytical model, the numerically predicted results of the height of workpiece are obtained. For the dome height of the hydroformed workpieces at each channel, the comparison of the analytical results and the experimental ones is shown in Table 5.1.

It can be seen that the analytical model is able to accurately predict the height of the workpiece hydroformed into channels with different geometric parameters. It further shows that the analytical model facilitates an in-depth understanding of the deformation behavior of thin sheet hydroformed into microchannel features. In addition, it is found that the results provided by the conventional method are much smaller than the experimental results and analytical ones. The main reason might be that the conventional model does not include the influence of the fillet, which can make the flow of the material easier, especially when the thickness of the sheet metal and the width of the channel are in the same magnitude. Therefore, the conventional model tends to "underestimate" the height of the deformed workpieces. According to the comparison of the analytical and experimental results, it is observed that the analytical results are smaller than the experimental ones at a low pressure, but they would be greater than the experimental results at high pressure. Springback may cause this tendency. On the other hand, friction is not included in the analytical model due to the fact that hydraulic oil is used as a pressure-carrying medium in the experiments, and the contact between the workpiece and the die is under good lubrication condition in the whole hydroforming process.

TABLE 5.1

Comparison of the Experimental and Analytical Results of the Dome Height [24]

Pressure (MPa)	Results	Channel 1	Channel 2	Channel 3	Channel 4	Channel 5
55	Experimental	0.269	0.66	0.317	0.185	0.488
	Analytical model	0.256	0.664	0.300	0.153	0.479
	Error%	−4.83	0.61	−5.36	−17.3	−1.84
	Conventional model	0.189	0.436	0.231	0.128	0.360
	Error%	−29.74	−33.94	−27.13	−30.81	−26.23
75	Experimental	0.298	0.774	0.376	0.222	0.586
	Analytical model	0.316	0.794	0.388	0.196	0.593
	Error%	6.04	2.58	3.19	−11.71	1.19
	Conventional model	0.246	0.551	0.302	0.172	0.465
	Error%	−17.45	−28.81	−19.68	−22.52	−20.65

5.5 Microflexible Forming of Sheet Metals by Soft Punch

Comparing with the traditional sheet forming process with rigid tools, meso- and microscale flexible forming of sheet metals by soft punch have many advantages. It only needs a rigid die, and the punch is made of soft material such as rubber. Therefore, only a rigid die must be designed and accurately manufactured, and it is also easy to accurately assemble the soft punch and rigid die. In this area, much prior research have been conducted. To name a few, Browne and Battikha [33] presented an experimental study of a rubber-forming process to produce sheet metal parts. They investigated the capability of the process and optimized the process parameters to ensure defect-free products to be made using a double-acting hydraulic press with the capacity of 100 tons. Sala [34] studied the significant parameters associated with the flexible forming process by numerical simulation with a commercially FE package. Their investigations showed that rubber hardness, blank material type, contact friction, and die design are crucial parameters that require adjustment before actual operations.

5.5.1 Interface Friction

In meso- and microforming of sheet metals using soft punch, there are two important interfaces, viz., the interface between rigid punch and metal sheet; and the interface between soft punch and metal sheet. To investigate the effect of friction between rigid punch and metal sheet, two rigid dies with different surface roughnesses were designed in the research done by Peng et al. [35] and shown in Figure 5.20. The experimental assembly for microforming of sheet metal by soft punch includes rigid punch, container, soft punch, metal sheet, rigid die, and die plate. For the two rigid dies, as shown in Figure 5.20b, one of them is a rough die and the other is a smooth die with good surface polishing. The surface roughness is measured, and the roughness values are 2.69 and 0.983 μm for the rough and smooth dies, respectively.

Based on the earlier designed die set, the stainless steel sheet of ASI304 with a thickness of 0.1 mm and a grain size of 25 μm was punched. The rubber with a Shore A hardness of 70 was used as the soft punch. The formed part under the load force of 80 kN is shown in Figure 5.21a. The formed parts were then cut in the middle along the dot line, and the thickness of the deformed microgroove parts was measured, respectively, at four different locations shown in Figure 5.21, middle of the bottom microgroove (O), the minimum thickness (C), then end of transition region (B), and the location 0.6 mm away from point *B* (*A*). Figure 5.21c shows the comparison of the numerical simulation results and those of the experimental ones. It is found that the thickness distribution of the deformed parts using rough die is more uneven than that of parts made by the smooth die, because it has a greater resistance to block the flow of material in the forming process.

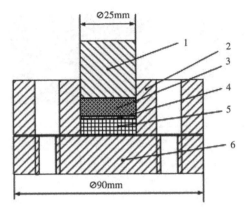

(a) Sketch of forming experiment assembly

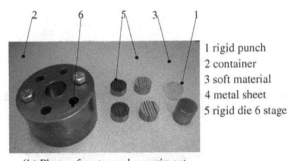

1 rigid punch
2 container
3 soft material
4 metal sheet
5 rigid die 6 stage

(b) Photo of parts used experiment

FIGURE 5.20
(a) Sketch of forming experiment assembly and (b) rough die and smooth die [35]. 1. Rigid punch, 2. die container, 3. soft material, 4. sheet metal, 5. rigid die, 6. die plate.

Different from the fiction between sheet metal and rigid die, which is a main parameter to influence the forming process, the friction between soft punch and sheet metal is not an important factor. Figure 5.22 shows the distribution of von Mises stress of the deformed sheets by soft punch with different friction conditions. The stress distributions are almost the same when the friction coefficient f_1 between the soft punch and sheet metal increases from 0 to 0.3, as shown in Figure 5.22a and b. However, the stress is increased and the stress distribution becomes more uneven when the frictional coefficient f_2 between the soft punch and sheet metal is increased from 0 to 0.3, as illustrated in Figure 5.22c and d. When f_2 is equal to 0.3, the stress concentration becomes remarkable, which means that the metal sheet is prone to be broken with the increase of friction between the sheet metal and rigid die.

In the flexible die formed by soft punch, the impact of the friction between soft punch and sheet metal is not so significant due to the fact that soft punch is deformed along with the metal sheet in the forming process. The relative motion in between is much smaller. In rigid die forming, the die does not

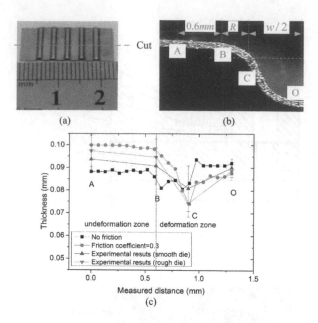

(a)   (b)

(c)

FIGURE 5.21
Thickness distribution in the section of the deformed parts [35] (a) formed parts (80 kN), (b) measure sketch, (c) thickness in the section of metal sheet.

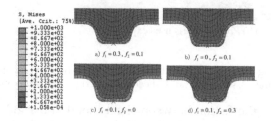

FIGURE 5.22
von Mises stress distribution of the deformed parts with different friction conditions [35]: (a) friction coefficient $f_1 = 0.3$ and $f_2 = 0.1$, (b) friction coefficient $f_1 = 0$ and $f_2 = 0.1$, (c) friction coefficient $f_1 = 0.1$ and $f_2 = 0$, (d) friction coefficient $f_1 = 0.1$ and $f_2 = 0.3$.

move and the relative motion in the interface is much bigger, so is the friction force, which makes the metal sheet more difficult to flow, especially when the friction coefficient is bigger.

5.5.2 Hardness of Soft Punch

In flexible meso- and microforming of sheet metals, flexible rubber materials are generally used as the soft punch. The impacts of soft punch on the deformation of sheet metals have been studied extensively. Currently, the most widely used soft material is polyurethane rubber. In the research done by Irthiea et al. [36], three types of polyurethane rubber materials with Shore A hardness of 40, 63, and 75 were used for the soft die. Through experiment and simulation, the thickness distribution along the transverse direction of the deformed cup and the comparison between the experiment and simulation is shown and presented in Figure 5.23. It is clear that the curves of the thickness distribution almost have the same tendency, where the maximum thinning value is located at the top portion of the cup between the flange and the vertical sidewall for all the deformed parts. The numerical prediction indicates that the maximum thinning is 17.4%, 15.8%, and 18% for the rubber materials with a Shore A hardness of 40, 63, and 75, respectively, while they are 19%, 15% and 19% for the experiments. These results indicate that by using different rubber materials for the soft die, the reduction of thickness is not much different in the formed cups.

To investigate the influence of soft punch's hardness, two kinds of rubbers with Shore A hardness of 55 and 70 are used in FE simulation. The Coulomb friction coefficients are defined as 0.1 for the two interfaces. In Figure 5.24, h and w are the height and the width of the channel. It is found that the final von Mises stress distribution is also the same in the stress-concentrated region. This means the hardness of the soft punch is not a decisive factor in micro flexible forming process. Though the hardness of soft punch is different, their capability to transfer the forming force is equal in a close container. That is why same results are shown in the microforming process.

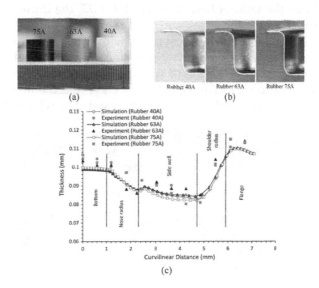

FIGURE 5.23
(a) Thickness distribution along transverse direction using different rubber materials [36]: (a) rubber punches (b) physical cups formed using different rubber materials, (c) thickness against curvilinear distance under various conditions.

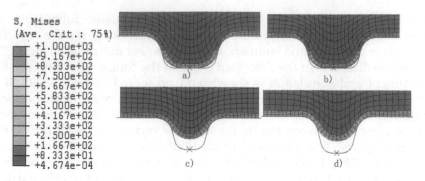

FIGURE 5.24
von Mises stress distribution of the formed sheet metal parts using the soft tools with different Shore A hardness: (a) Shore A hardness 70 ($h/w = 0.5$), (b) Shore A hardness 50 ($h/w = 0.5$), (c) Shore A hardness 70 ($h/w = 0.75$), (d) Shore A hardness 50 ($h/w = 0.75$) [35].

5.5.3 Grain Size Effect

The grain size of sheet metals affects the deformation behavior in meso- and microforming process by soft punch. To explore this, Peng et al. [35] conducted a research on the effect of grain size using a stainless steel sheet with various grain sizes by FE simulation. The simulations were done with the same lubricant condition and soft punch with a hardness of 70. From Figure 5.25, it can be seen that the stress distribution of the sheet metal with the grain size of 25 µm is more even than that of the sheets with the grain size of 100 µm. When the ratio of height to width is 0.75, the draw depth is 0.04, 0.044, and 0.046 mm for the cases with the grain size of 25, 60, and 100 µm, respectively. That is because the sheet metal with smaller grains has a bigger resistant force to the deformation; therefore, the drawn channel is shallower.

The final thickness distribution of the half-formed sheet parts with various grain sizes is shown in Figure 5.26. It is clear that the most dangerous region is located at the bell bottom of the rigid die, where the stress is

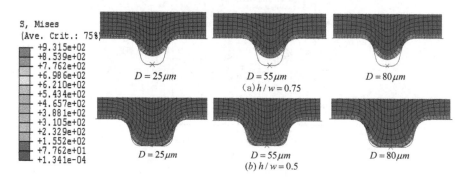

FIGURE 5.25
Final shape of formed parts with different grain sizes: (a) final shape of formed parts ($w = 0.8$, $h/w = 0.75$) and (b) final shape of formed parts ($w = 1.2$, $h/w = 0.5$) [35].

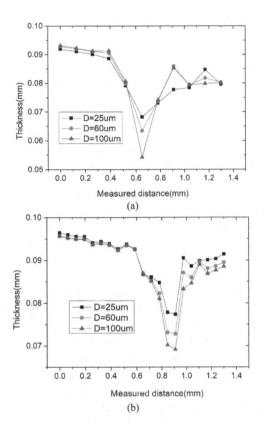

FIGURE 5.26
Final thickness of the sheets at various grain and feature sizes [35] (a) final thickness distribution ($w = 1.2$, $h/w = 0.5$), (b) final thickness distribution ($w = 0.8$, $h/w = 0.75$).

concentrated, to make the severe deformation and finally the crack of sheet metals. For the sheet with $h/w = 0.5$ and the original thickness of 0.1 mm, the thickness is reduced to 0.054 mm for the sheet with the grain size of 100 μm; 0.063 mm for the sheet with the grain size of 60 μm; and 0.068 mm for the case with the grain size of 25 μm. The thickness distribution for the sheet with the grain size of 25 μm is much more even than that of the sheet with the grain size of 100 μm, as shown in in Figure 5.26a. The same conclusion can be obtained from the simulation results of the sheet with $h/w = 0.75$, as shown in Figure 5.26b. It is found that most deformation is located within the microgroove, whereas the material outside the bell of the rigid die is much less deformed in the forming process. Also, the sheet metal with the larger grains is easier to crack in the forming process. The reason is that the sheet metal with smaller grains has more grains located at the section so that more grains contribute to the limited deformation than the sheets with larger grains.

5.6 Summary

Meso- and microsheet metal forming provide an economic and promising approach for production of miniaturized metallic parts. However, the scaling down of the parts to be made from macro- to meso- and microscale leads to the occurrence of different size effects and the size-affected deformation behaviors in the forming process. In this chapter, the meso- and microforming processes of sheet metals, including meso- and microbending, deep drawing, hydroforming, and sheet metal flexible forming by soft punch are represented.

In meso- and microsheet metal bending, the first stage shown in the load–displacement curve is the three-point bending stage, in which the forming force is increased gradually with the punch angle of specimen as the punch moves down. The second one is the sliding stage. Once the punch angle reaches a critical value, the specimen gradually slides into the groove. Upon deformation, the springback of sheet metal occurs, which is another eluded issue attracting lots of attention. It is found that the springback angle in the bending test is increased with the decrease of thickness for the specimens with a similar grain size, and is increased with the increase of t/d for the specimens with the same thickness. Furthermore, with the same value of t/d, the springback behavior of specimens with greater thickness is also found to be less obvious.

The deformation load of the multistage drawing in meso- and microscale is represented in this chapter. In the first stage, the deformation load is increased first to the maximum point, and then decreased to form the first peak. When the thick flange is drawn into the clearance between the punch and die, the second peak occurs. In the second stage, there is no obvious peak in the load–stroke curve. In addition, the thickness variation of microparts is discussed. For the copper annealed at 750°C, nonuniform thickness distribution and severe thinning around the punch corner of the microparts are more significant. The maximum thickness thinning is approximately 25% and 35% for the microparts drawn by the blank annealed at 750°C in the first stage and second stage, respectively, which are larger than that using the blank annealed at 500°C.

By meso- and microsheet hydroforming process, the microchannels with different widths from 0.8 to 1.6 mm, radius of the fillet with the size range from 0.2 to 0.6 mm, and the height from 0.4 to 1.2 mm, were successfully formed using the SS304 sheets with a thickness of 0.1 mm. The deformation of sheet shows two stages. In Stage I, the sheet begins to flow into the channel and contacts the fillet; In Stage II, the sheet is bulged into the channel and contacts the inner walls of the channel. Correspondingly, the deformed workpiece can be divided into three areas according to different loading conditions during the forming process: free bulging area, fillet area, and flat area. A good prediction of the deformation of microchannels in microhydroforming process can be done by analysis approach.

In meso- and microsheet flexible forming by soft punch, the influence of the friction between metal sheet and rigid is much more evident than that of the friction between soft punch and metal sheet. This is caused by the fact that the relative motion and the friction influence between soft punch and metal sheet are much smaller. The hardness of the soft punch is not a decisive factor, since the capabilities of the rubbers to transfer the forming force are almost the same in a close container. In addition, the sheet metal with smaller grains has a bigger resistance to deformation and the corresponding draw channel is thus shallower. The sheet metal with larger grains is easier to crack in the forming process.

References

1. K. Suzuki, Y. Matsuki, K. Masaki, M. Sato, M. Kuroda, Tensile and microbend tests of pure aluminum foils with different thicknesses, *Materials Science and Engineering: A* 513 (2009) 77–82.
2. J.G. Liu, M.W. Fu, J. Lu, W.L. Chan, Influence of size effect on the springback of sheet metal foils in micro-bending, *Computational Materials Science* 50(9) (2011) 2604–2614.
3. J. Wang, M. Fu, J. Ran, Analysis of the size effect on springback behavior in micro-scaled U-bending process of sheet metals, *Advanced Engineering Materials* 16(4) (2014) 421–432.
4. Z. Xu, L. Peng, E. Bao, Size effect affected springback in micro/meso scale bending process: Experiments and numerical modeling, *Journal of Materials Processing Technology* 252 (2018) 407–420.
5. Y. Saotome, K. Yasuda, H. Kaga, Microdeep drawability of very thin sheet steels, *Journal of Materials Processing Technology* 113(1–3) (2001) 641–647.
6. F. Vollertsen, Effects on the deep drawing diagram in micro forming, *Production Engineering* 6(1) (2012) 11–18.
7. T. Shimizu, H. Komiya, T. Watanabe, Y. Teranishi, H. Nagasaka, K. Morikawa, M. Yang, HIPIMS deposition of TiAlN films on inner wall of micro-dies and its applicability in micro-sheet metal forming, *Surface and Coatings Technology* 250 (2014) 44–51.
8. H. Tanabe, M. Yang, Design and evaluation of heat assisted microforming system, *Steel Research International Special* 81(9) (2011) 1020–1024.
9. K. Manabe, T. Shimizu, H. Koyama, M. Yang, K. Ito, Validation of FE simulation based on surface roughness model in micro-deep drawing, *Journal of Materials Processing Technology* 204(1) (2008) 89–93.
10. I. Irthiea, G. Green, S. Hashim, A. Kriama, Experimental and numerical investigation on micro deep drawing process of stainless steel 304 foil using flexible tools, *International Journal of Machine Tools and Manufacture* 76 (2014) 21–33.
11. F. Vollertsen, D. Biermann, H.N. Hansen, I.S. Jawahir, K. Kuzman, Size effects in manufacturing of metallic components, *CIRP Annals – Manufacturing Technology* 58(2) (2009) 566–587.

12. Y.M. Huang, Y.S. Wu, J.Y. Huang, The influence of ultrasonic vibration-assisted micro-deep drawing process, *The International Journal of Advanced Manufacturing Technology* 71(5) (2014) 1455–1461.

13. W.T. Li, M.W. Fu, J.L. Wang, B. Meng, Grain size effect on multi-stage micro deep drawing of micro cup with domed bottom, *International Journal of Precision Engineering and Manufacturing* 17(6) (2016) 765–773.

14. M. Ahmetoglu, T. Altan, Tube hydroforming: State-of-the-art and future trends, *Journal of Materials Processing Technology* 98(1) (2000) 25–33.

15. M. Koç, T. Altan, An overall review of the tube hydroforming (THF) technology, *Journal of Materials Processing Technology* 108(3) (2001) 384–393.

16. S.H. Zhang, Z.R. Wang, Y. Xu, Z.T. Wang, L.X. Zhou, Recent developments in sheet hydroforming technology, *Journal of Materials Processing Technology* 151(1) (2004) 237–241.

17. P. Hein, F. Vollertsen, Hydroforming of sheet metal pairs, *Journal of Materials Processing Technology* 87(1–3) (1999) 154–164.

18. O. Kreis, P. Hein, Manufacturing system for the integrated hydroforming, trimming and welding of sheet metal pairs, *Journal of Materials Processing Technology* 115(1) (2001) 49–54.

19. S. Novotny, P. Hein Hydroforming of sheet metal pairs from aluminium alloys, *Journal of Materials Processing Technology* 115(1) (2001) 65–69.

20. M. Ahmetoglu, J. Hua, S. Kulukuru, T. Altan, Hydroforming of sheet metal using a viscous pressure medium, *Journal of Materials Processing Technology* 146(1) (2004) 97–107.

21. B.Y. Joo, S.I. Oh, Y.K. Son, Forming of micro channels with ultra thin metal foils, *Cirp Annals-Manufacturing Technology* 53(1) (2004) 243–246.

22. S. Mahabunphachai, M. Koç, Fabrication of micro-channel arrays on thin metallic sheet using internal fluid pressure: Investigations on size effects and development of design guidelines, *Journal of Power Sources* 175(1) (2008) 363–371.

23. L.F. Peng, P. Hu, X.M. Lai, D.Q. Mei, J. Ni, Investigation of micro/meso sheet soft punch stamping process–Simulation and experiments, *Materials & Design* 30(3) (2009) 783–790.

24. Z. Xu, L. Peng, P. Yi, X. Lai, Modeling of microchannel hydroforming process with thin metallic sheets, *Journal of Engineering Materials and Technology* 134(2) (2012) 021017.

25. R. Hill, A theory of the plastic bulging of a metal diaphragm by lateral pressure, *Philosophical Magazine* 41(322) (1950) 1133–1142.

26. J. Chakrabarty, J. Alexander, Hydrostatic bulging of circular diaphragms, *The Journal of Strain Analysis for Engineering Design* 5(3) (1970) 155–161.

27. G. Gutscher, H.C. Wu, G. Ngaile, T. Altan, Determination of flow stress for sheet metal forming using the viscous pressure bulge (VPB) test, *Journal of Materials Processing Technology* 146(1) (2004) 1–7.

28. H.M. Shang, S. Qin, C.J. Tay, Hydroforming sheet metal with intermittent changes in the draw-in condition of the flange, *Journal of Materials Processing Technology* 63(1–3) (1997) 72–76.

29. H. Shang, F. Chau, K. Lee, C. Tay, S. Toh, Modeling of the hydroforming of sheet materials clamped with varying blank holding loads, *Journal of Engineering Materials and Technology* 109 (1987) 92.

30. A.A. Kruglov, F.U. Enikeev, R.Y. Lutfullin, Superplastic forming of a spherical shell out a welded envelope, *Materials Science and Engineering: A* 323(1–2) (2002) 416–426.

31. A. Assempour, M.R. Emami, Pressure estimation in the hydroforming process of sheet metal pairs with the method of upper bound analysis, *Journal of Materials Processing Technology* 209(5) (2009) 2270–2276.
32. M. Koç, E. Billur, Ö.N. Cora, An experimental study on the comparative assessment of hydraulic bulge test analysis methods, *Materials & Design* 32(1) (2011) 272–281.
33. D.J. Browne, E. Battikha, Optimisation of aluminium sheet forming using a flexible die, *Journal of Materials Processing Technology* 55(3) (1995) 218–223.
34. G. Sala, A numerical and experimental approach to optimise sheet stamping technologies: Part II—Aluminium alloys rubber-forming, *Materials & Design* 22(4) (2001) 299–315.
35. L. Peng, P. Hu, X. Lai, D. Mei, J. Ni, Investigation of micro/meso sheet soft punch stamping process—Simulation and experiments, *Materials & Design* 30(3) (2009) 783–790.
36. I.K. Irthiea, G. Green, Evaluation of micro deep drawing technique using soft die-simulation and experiments, *The International Journal of Advanced Manufacturing Technology* 89(5) (2017) 2363–2374.

6

Meso- and Microscaled Sheet-Metal Parts Made by Meso- and Microstamping Processes

6.1 Introduction

With the increasing demand for high-quality miniaturized products, such as portable consumer electronics and other meso- and microscaled devices, and the sheet metal parts and components with characterized and localized microscaled features, such as microchannel bipolar plate (BPP) in fuel cells, micromanufacturing technologies are critically needed. Microforming, as one of the promising micromanufacturing technologies, has achieved remarkable development and has been used in some relevant industries in the past few decades. The meso- and microforming processes including meso- and microscaled stamping, embossing, punching, coining, extrusion, and progressive and compound bulk forming have been explored and developed [1–4]. Among these approaches, meso- and microstamping has been well studied for their promising potential applications in mass production of micrometallic products in different industrial clusters for its low production cost, high efficiency, limited space occupation, and low energy consumption. Due to these advantages, meso- and microstamping process has been extensively used to produce miniaturized parts and components in different arseas such as microelectromechanical systems, renewable energy, electronics, biomedicines, aerospace, microreactors, etc.

Until now, numerous research projects and investigations have been conducted on meso- and microstamping, and a considerable improvement in process capability has been done in recent years in terms of formation of new materials, development of new products, and viability of forming devices and processes. These studies indicate that meso- and microstamping has a promising potential for large-scale applications in industries. Therefore, extra efforts have been provided to explore the fully automatic meso- and microstamping for the production of miniaturized products. However, there is a big gap between research conducted in laboratories and actual production. Some critical issues should be successfully addressed before the wide application of unique meso- and microforming processes.

The first issue is the design of microproducts. Design of microproducts for meso- and microstamping needs to consider the production issues, viz., design for manufacturing, such that the design can be efficiently fabricated and the capability of the process can be fully employed and considered in upfront design stage [5]. For macromanufacturing processes, such as milling, the requirement for considering the material formability is not so critical in the design stage. For meso- and microforming, achieving the desirable microproduct geometries and dimensional accuracy would be a common and serious issue if a complex design is employed, leading to the difficulty in production. By using metallic BPP as a fairly typical meso- and microstamping case, the BPP produced by this process is a good choice for replacing the traditional carbon-based graphite BPP of a proton exchange membrane (PEM) fuel cell. BPP, as the skeleton of the fuel cell, is formed with many meso- and microscaled dense features to distribute reactant gases and manage the water in working of the fuel cell. However, fractures are often observed in meso- and microforming processes if the design guidelines of the graphite BPP traditionally fabricated by machining are simply employed to support the design of the thin metallic BPP. Reactant leakage or short circuit would inevitably occur when the BPP with microfractures is assembled into the fuel cell. In addition, the design should extensively address several avoidable forming issues in the deformation process, such as the avoidable radius of stamping tools, the tooling construction capabilities, the constraints among the neighboring meso- and microfeatures, and the dimensional accuracy of the metal sheet formed parts. All of these factors pose new challenges to the design of meso- and microproducts by stamping process. It is thus critical to simultaneously consider the functional requirements and the formability of the metallic materials in production, and hence an optimal tradeoff can be made.

The second issue is the size effect in meso- and microforming. When the size of the mechanical parts is scaled down to less than 1 mm, the so-called size effects occur, which make the traditional knowhow, empirical and analytical methods in the conventional macroforming processes unsuitable for microforming. Size effect has been known to exist in any microforming processes, such as stamping [6], bending [7], forming [8], molding [9], and forward and backward extrusion [10]. As a result, the production parameters in macroscale cannot be simply scaled down according to the scaling law for their applications in microscale. The flow stress, grain size effect, interfacial friction, anisotropy, ductility, and forming limit of materials are also influenced by the size effect and change. These factors should be reevaluated, revisited, and revised for their applications in meso- and microforming [11]. In addition, some factors that can be ignored in the conventional macroforming may play a bigger role in meso- and microstamping processes, such as vibration, part weight, tool offset, and working temperature. Since the forming features of the parts are in meso- and microscale, even a small tolerance with a few microns would likely induce a detrimental influence on the accuracy of microproducts. Therefore, the basic forming theory, knowhow, and more robust process strategies should be improved to make the meso- and microforming more viable and stable.

The third issue to be addressed is the precision of meso- and microstamped products in the complex-forming process. The stamping process includes two separated procedures. One is the fabrication of the microstamping tools (punch, die, etc.), while the other is to use the developed tools to stamp the sheet metal parts with various microfeatures onto the thin sheet via plastic deformation. First, the tools should be meticulously designed and fabricated by precision manufacturing processes to ensure well-defined structures of relief on the tooling surfaces such as computer numerical control (CNC) machining, microwire electric discharge machining (EDM), laser ablation, imprinting, etc. The thin metal sheet is generally fed and positioned to the forming tools in the press working continuously or intermittently. Fabricating the stamping tools and feeding and positioning the metal sheet are generally not critical issues in the conventional production, while the challenges arise if the accuracies of tool making and feeding and positioning of sheet metal are less than the submicrometer range. If so, it is very difficult, if not impossible, to achieve high-quality deformed features with submillimeter size. In addition, the design parameters, such as clearance, blank holder, drawbead radius, and selection of lubrication, are the main concerns in the detailed design of the tools. Furthermore, the interactions between these factors would induce a number of complex issues to be addressed. Second, stamping with the optimized process parameters is conducted to make quality parts in terms of the property and quality of the deformed parts. From property and quality perspectives, the forming process involves the synthesis of deformation behavior and process parameters such as stamping rate, depth, and friction. Material thinning, springback, wrinkle, and even fracture are prone to occur in meso- and microstamping processes, leading to poor forming precision and defective products. Evaluating and controlling these avoidable deficiencies to a certain acceptable level is still a nontrivial issue. Therefore, tool design and fabrication, stamping operation, and prediction of deformation behavior and quality control in meso- and microforming processes should be performed with sufficient information and knowledge to support, in such a way high-quality miniature products can be designed and developed.

The chapter thus aims at illustrating the applications of meso- and microstamping technologies in industries, and the detailed design and development activities are systematically described and presented. To facilitate detailed presentation, some basic formability theories of meso- and microstamping processes are also introduced and summarized.

6.2 Meso- and Microsheet Metal Products Made by Stamping Processes

Meso- and microstamping technology offers a promising approach for a large-scale production of meso- and microproducts or the macroscale products with microscale features, such as chemical reaction, heat exchange application, mixing, and flow distribution. Some novel applications of meso- and microstamping processes are given in the following.

6.2.1 Microreactor

Microreactors, evolved from the process intensification concept and the microfabrication techniques, create a promising potential for high-performance chemical and information processing due to the considerable surface-to-volume ratio [12]. In recent years, microreactors have been used in many fields, such as methane steam reforming, fuel cell, hydrogen peroxide production, and gas-to-liquid (GTL) technology. In the last two decades, powerful fabrication technologies have been developed to fabricate microreactors from a wide variety of materials, such as photochemical etching for monocrystalline materials, anisotropic etching for photosensitive glass, and the traditional milling, laser cutting, and stamping. For sheet metal reactors, meso- and microstamping is considered the most economical technique for mass production. Figure 6.1 shows several microchannel reactors used in hydrogen production, GTL reactor, and fuel cell [13–15]. In these applications, meso- and microfeatures are fabricated on the metal sheets in a designed pattern to create the desirable flow network for gases or liquid in the microreactor. To improve the working stability of reactors, surface modification technique is also employed for meso- and microformed sheet metal components.

The microstructures on the components are important to the efficiency of reactors. Reuse et al. [13] fabricated the "S"-shaped channels with a width of 320 μm and a depth of 100 μm onto the aluminum with the thickness of 200 μm, and a two-passage microstructured reactor was developed for the coupled functions of methanol steam reforming and total oxidation reaction. Ehrfeld et al. [16] developed a catalyst structure comprising an array of microchannels with a width of 60 μm for hydrogen cyanide (HCN) generation. In addition, the GTL process is used to convert the gas resource to liquid product, which is cheaper and easier to be transported in the marketplace. With the advent of meso- and microforming technology, the current microreactors extend the capability range of the conventional GTL reactors for faster reaction due to the enhanced mass and heat transfer. They are improved by the microchannels with a typical width of 300 μm, a depth of 100 μm, and a length of 78 mm separated by the wide walls of 100 μm, which are fabricated by the University College London and Institut für Mikrotechnik

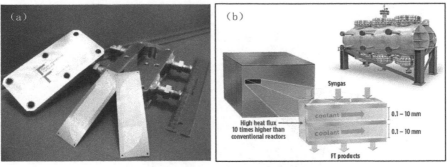

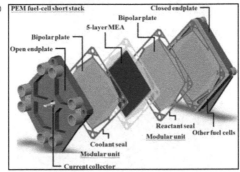

FIGURE 6.1
(a) Microstructured reactor for hydrogen production [13], (b) the microchannels (left) in the GTL reactor and the reactor in a full-pressure shell (right) [14], and (c) fuel cell with microchannel BPP [15].

Mainz GmbH [17,18]. It should be noted that most of earlier microchannel reactors are still at the research and development stage. Comparatively, with the development of new energy devices, the meso- and microstamping technology has been widely used to fabricate the metallic BPP of fuel cell [15]. The launched fuel cell cars of Toyota, Honda, and Hyundai Motor utilized metallic BPPs to replace the conventional carbon-based BPPs to improve the power density and the mechanical strength of the fuel cell. The metal sheets with a thickness of 0.1–0.2mm are commonly used as BPP materials, and the formed features are typical with a width of 0.8–2.0mm and a depth of 0.4–1.0mm.

Most of the previous studies are focused on reactant distribution, catalyst incorporation techniques on channel surface, and the two-phase flow to improve the reaction efficiency of meso- and microreactors. However, with the requirement of high quality and excessive-design structures, many of the forming problems of meso- and microproducts arise when the desired geometries are stamped onto the thin metal sheets. Figure 6.2 shows several typical forming problems in meso- and microstamping processes [21–23]. First, the geometry error occurs due to elastic recovery, and the designed structures are difficult to achieve. The formed micro and dense features on

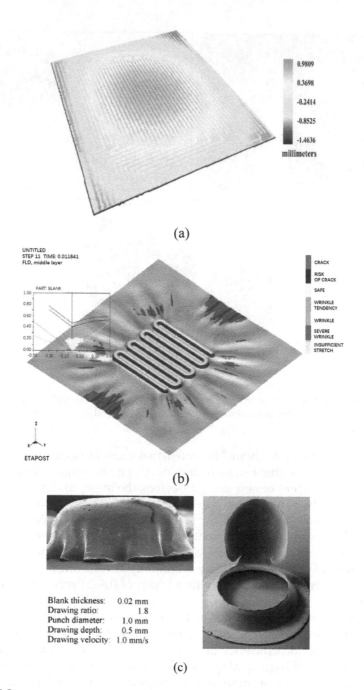

FIGURE 6.2
Typical forming problems in meso- and microstamping processes: (a) the measured shape deformation of metallic BPP with parallel channels [21], (b) the wrinkle formation of metallic sheet [22], and (c) the crack of thin metallic sheet [23].

the thin sheet are formed with various geometries. For example, the heights of multiple parallel channels (designed with a width of 1.0 mm and a depth of 0.4 mm) are decreased linearly from the edge channel to the center channel on the metallic BPP [19]. This was likely to be caused by the higher constraint of the material in the middle of the product. The height variation of 4.1% was also found on Mahabunphachai et al.'s parallel-channel samples [20]. Second, it is hard to keep the sheet metal products in flat shape after the stamping force is removed and springback, as shown in Figure 6.2a. Keeping the flat shape of macroproducts is probably not a critical issue, but it poses a challenge for the formed ultrathin sheet metal parts. The wrinkle and fracture of the thin sheet metal parts, as shown in Figure 6.2b and c, happen due to the nonuniform deformation of the large-area metal sheet and the localized thinning of materials. The poor material flow ability leads to the difficulty in fabricating the not well-designed features such as deep channel. In addition, the uniformity of material properties, the handling accuracy of stamping tools and the evenness of stamping force may contribute to these problems.

6.2.2 Micro Heat Exchanger

Compared with the conventional-channel ($\geq$6 mm) heat exchanger, the microchannel ($\leq$1 mm) heat exchanger has advantages such as higher heat exchange efficiency and less usage of working medium. In addition, its heat transfer coefficient can be improved with the reduction of channel geometries. Since the society pays more attention to energy conservation, the microchannel heat exchanger has thus been widely accepted in air-conditioning, automobile, consumer electronics, industrial clusters, etc. Figure 6.3 shows the microchannel heat exchangers [24–26].

Earlier, researchers [27] have found out that microchannel heat exchangers possess a weakness of low production rate, high fabrication cost, and lack of "accurate performance prediction tools," which hinder their commercial deployment and large-scale applications. Owing to the advance of meso- and microforming techniques in recent years, the meso- and microchannel heat exchangers have gained wide applications and have been regarded as a disruptive technology for heat exchange. Generally, heat exchangers can be categorized into metallic, silicon, polymer, and ceramic exchangers in terms of the materials used. For the metallic heat exchanger, the functional structures can be formed using the well-known manufacturing techniques, such as precision cutting with the profiled microdiamonds, multitool milling, sintering, chemical etching, and stamping. The stamping process is the preferred choice for the metallic plate heat exchanger. By using the meso- and microstamping processes, a lot of heat exchanger products have been produced. Naming a few cases, Tonkovich et al. developed a heat exchanger with the laminated microchannels [28]. Cho et al. also constructed a heat

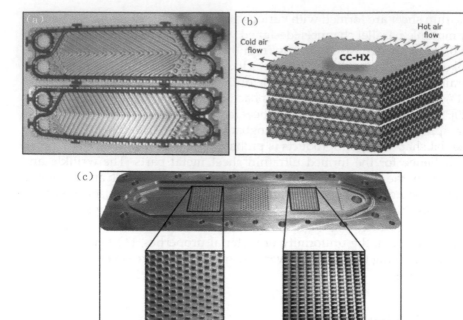

FIGURE 6.3
Microchannel heat exchangers: (a) axial-dispersion heat exchanger [24], (b) cross-corrugated heat exchanger [25], and (c) discrete-channel heat exchanger [26].

exchanger with a cross corrugated for aeroengines [29], and Ahn et al. came out with a stacked cross-corrugated exchanger [25].

To design the heat exchanger, it is essential to guarantee the uniform cooling and surface temperature, which require the same flow resistance in all the channel paths. To obtain a uniform flow distribution, researchers have done a series of investigations on manifold layout, phase of flow, Reynolds number, secondary distributor system, and channel dimensions [30]. Among these parameters, channel dimension is demonstrated to be the dominating factor for flow distribution. A high aspect ratio results in a more uniform flow distribution. However, the channels in micro heat exchangers are different from the conventional channels in terms of channel hydraulic diameters. For liquids, it is found that the smaller channel height is better for a good heat transfer efficiency with the intermediate velocity, while the larger channel height is better for higher and lower velocities [31]. Abdel-Salam et al. [32] reported that increasing manufacturing tolerance in channel spacing and decreasing the width resulted in a decreasing performance of the heat exchanger. An air channel thickness of 2 mm and the standard deviation variation from 1.2 to 1.8 mm made the effectiveness to be reduced from 49% to 37%. While for the channel with a thickness of 6 mm, only 1.7% drop was found for its effectiveness. Hence, channel dimensions and its accuracy should be carefully considered in the meso- and microforming processes.

6.2.3 Micromixer

Micromixer is used to mix several fluids into one main stream by collection of the multiple slit-shaped features in a microfluidic system, which can be used in mechanical machines, biochemistry, drug delivery, etc. [33]. In a micromixer, the dispersion of fluids caused by small size and narrow-distribution features offers advantages of uniform and quick mixing. Previously, the application of micromixer is mainly in the electromechanical systems. Although the current mixers can be made of new materials, such as silicon, glass, and polymer, metallic micromixer, as shown in Figure 6.4, is still an important and popular one due to its high mechanical strength and stain resistance [34,35].

Compared with the conventional mixer, the effective mixing of fluids within microfluidic devices is challenging due to their low Reynold number flow conditions [36]. Most of the current studies aim to improve the mixing efficiency by channel surface modification and optimization of the structure configuration and channel geometries. For the silicon and polymeric mixers

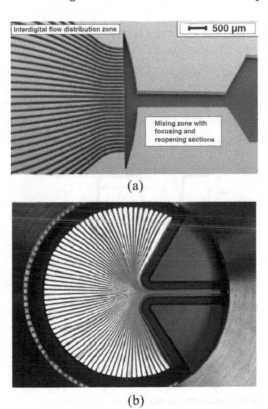

(a)

(b)

FIGURE 6.4
(a) Slit-type interdigital micromixer for high volume flow [34]; (b) 138 microchannel liquid lamellae in a mixing chamber of SuperFocus micromixing [35].

fabricated by ion etching, hot embossing, and layered manufacturing, the manufacturability of any design may not be a critical issue. For meso- and microstamping process, there are some limitations. This approach does have more advantages such as low manufacturing cost, especially for the two-dimensional channel structure, which is proven to be a simple and promising method for mass production [37].

6.3 Process Realization: A Case Study

To present the main concerns in applications of meso- and microstamping process, the metallic BPP of fuel cell fabricated by meso- and microforming processes are used as a case study in this chapter. Meso- and microstamping with rigid and soft punches are employed to fabricate metallic BPPs, respectively.

6.3.1 Process Realization and the Issues Involved in Fabrication of BPPs

A typical PEM fuel cell is mainly composed of membrane exchange assembly and BPP, as shown in Figure 6.5. Traditionally, the carbon-based BPP accounts for about 60%–80% of stack weight and 30%–45% of stack cost. In recent years, the metallic BPPs have begun to replace the traditional carbon-based BPPs of fuel cell for its good mechanical strength, thinness,

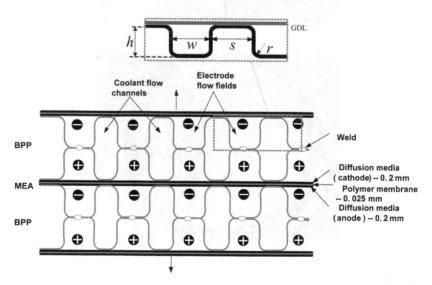

FIGURE 6.5
Structure of the fuel cell with metallic BPPs.

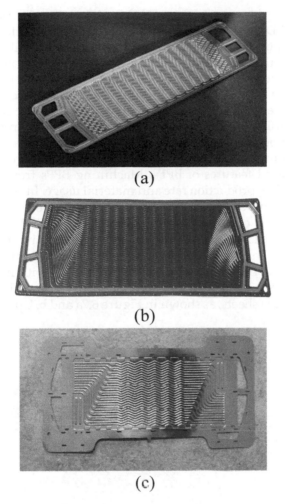

FIGURE 6.6
Typical metallic BPPs: (a) metallic BPP by Borit Inc. [39]; (b) metallic BPP by SJTU-Shanghai Zhizhen Inc. [40]; (c) metallic BPP by Tallistech Inc. [41].

good productivity, and low cost in mass production. Figure 6.6 shows the commercial metallic BPPs produced by several companies [39–41]. The flow channels formed on BPPs are in microscale ($\approx$1 mm) and play an important role in stack operation, such as (1) providing independent reaction channels for fuels and oxygen, (2) exporting excess water from stack, (3) making an electric conduction between adjacent fuel cells [19,38].

Over the past two decades, there is plenty of work focusing on the design and manufacture of metallic BPP. Various approaches, as shown in Figure 6.7, were proposed to fabricate the crucial components of the fuel cell, such as meso- and microstamping [42,43], hydroforming [44], electrochemical micromachining (EMM) [45], microelectrical discharge machining (µEDM) [46], rolling [47], and electromagnetic forming [48]. However, most of these fabrication technologies have limitations in terms of mass production of the parts. Taking hydroforming as an instance, the cost of the facility is pretty expensive, and liquid sealing is also a critical issue due to the high hydroforming pressure. The metallic BPP fabricated by electromagnetic forming process with electromagnetic coils and forming dies has difficulty in obtaining the needed channel depth, and the coil service life is not long enough and sufficient to commercialize the electromagnetic forming process. Rolling process has limitations in the formation of complexed features of BPPs. Machining BPPs from a thick metal block involves low production rate and material usage. In addition, the productivity using these technologies for making BPPs needs to be enhanced further. Therefore, considering the requirements of BPPs manufacturing in terms of precision, production rate, and cost effectiveness, meso- and microstamping can be regarded as an excellent choice for the manufacture of metallic BPPs [49]. In this process, the rigid dies with micromachined channels are pressed by a rigid punch or soft punch to deform the thin stainless steel (SS) sheets, as shown in Figure 6.7a and b. Upon deformation,

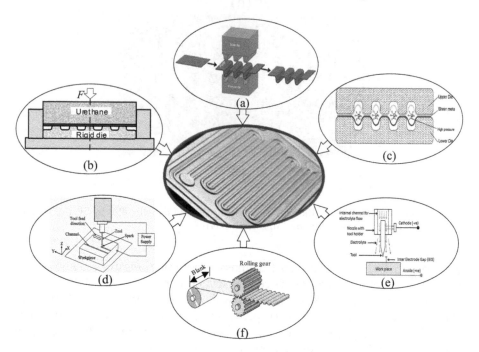

FIGURE 6.7
The main forming processes of metallic BPP: (a) rigid stamping [42]; (b) soft stamping; (c) hydroforming [44]; (d) µEDM [50]; (e) EMM [51]; and (f) rolling.

hundreds of microchannels are embossed on each side of the BPPs, and in this way, BPPs are produced. By using the meso- and microstamping process for fabrication of BPPs, the unique characters such as low cost and high productivity are outstandingly highlighted.

The design of metallic BPPs has posed more challenges than the conventional product design due to the characteristics of meso- and microforming employed for making BPPs. For the conventional carbon-based BPPs, plenty of configurations and section shapes of the flow channels were proposed and developed to address some technical issues such as uneven flow distribution and water flooding during the cell operation [52,53]. Those design guidelines on flow channels do not cover the manufacturability of BPPs since greater thickness is adopted for carbon-based BPP in machining or injection molding due to material brittleness. Nevertheless, due to small thickness and stress concentration induced in the deformation process, the excessive channel shape design of the metallic BPP results in the higher risk of rapture during the sheet meso- and microstamping process, and the difficulty in making from a manufacturability perspective [54,55]. Therefore, the design guidelines for carbon-based BPPs are not fully valid for thin metallic BPPs. Like other meso- and microproducts, it is necessary to investigate design for manufacturing, and whether the designed geometries can be formed by meso- and microstamping process is more critical for small-scaled products.

In actual production, a BPP specimen is usually fabricated to observe formability of the metallic sheet and feasibility of the desirable design. As shown in Figure 6.8, the specimen is designed by multiple types of channels with various dimensions. After the tooling fabrication and stamping of the specimen, the formed features can be measured to provide a guideline for the real design of the metallic BPP. However, this method is quite costly in both monetary and temporal terms. To achieve more effective results, the analytical approach and finite element (FE) method are proposed and used in this section.

6.3.2 Meso- and Microstamping with Rigid Punch

6.3.2.1 Rigid Tooling Stamping

The formability of microchannels in the rigid tooling stamping is investigated based on the developed instability criterion. Figure 6.9 shows the deep drawing of the metallic sheet in manufacturing metallic BPPs. As shown in Figure 6.9, it is not easy to determine the strain during the stamping process; however, the punch stroke can be easily measured. On the other hand, the forming limit from the angle of depth is a key concern in channel design. Therefore, the relationship between the strain and forming depth is established using the following analysis.

6.3.2.1.1 Force Balance Analysis

As shown in Figure 6.9, fracture is more likely to occur at the central channel due to the constraints such as punch and die. Since the channel has symmetrical configuration, half of the central channel, divided into five parts as shown in Figure 6.10, is taken into account in the analysis.

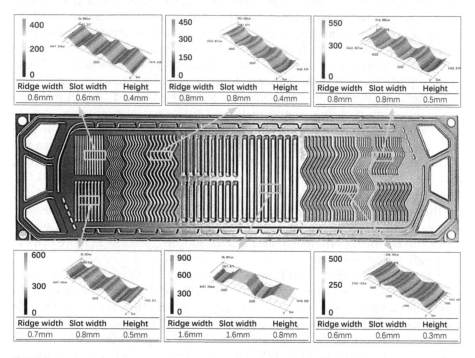

Ridge width	Slot width	Height
0.6mm	0.6mm	0.4mm

Ridge width	Slot width	Height
0.8mm	0.8mm	0.4mm

Ridge width	Slot width	Height
0.8mm	0.8mm	0.5mm

Ridge width	Slot width	Height
0.7mm	0.8mm	0.5mm

Ridge width	Slot width	Height
1.6mm	1.6mm	0.8mm

Ridge width	Slot width	Height
0.6mm	0.6mm	0.3mm

FIGURE 6.8
A BPP specimen fabricated with various channel geometries.

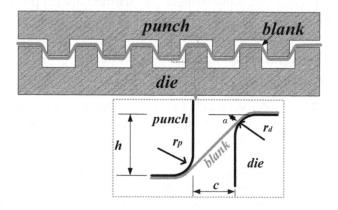

FIGURE 6.9
Schematic diagram of the deep drawing process (r_p – punch radius, r_d – die radius, c – clearance, α – wrap angle).

For AB section, it is equilibrated under only two tensions T_A, T_B and $T_A = T_B$. For Part BC with the punch radius, the element of arc length illustrated in Figure 6.11a is selected to be analyzed and shown in Figure 6.11.

The equilibrium equation of the forces along the sheet is

$$\frac{dT_1}{T_1} = \mu' d\theta \tag{6.1}$$

Integrating Eq. (6.1), we get the following:

$$T_C = T_B e^{\mu' \alpha} \tag{6.2}$$

where α is the so-called wrap angle, as shown in Figure 6.9.

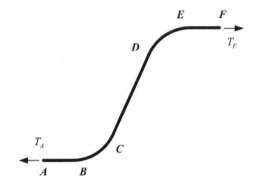

FIGURE 6.10
Analysis diagram of the forming blank.

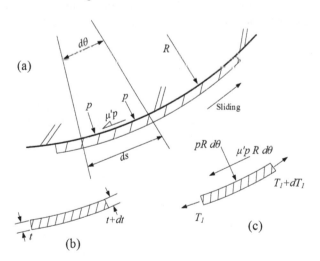

FIGURE 6.11
Force analysis of the element on the corner of the punch: (a) force status at the radius of punch, (b) thickness distribution, and (c) element force status.

Considering the force balance of other parts in the sheet, the following are established:

$$\begin{cases} T_C = T_D \\ T_E = T_D e^{-\mu\alpha} \\ T_F = T_E \\ T_A = T_F \end{cases} \tag{6.3}$$

The force balance in the stamping process is built by Eqs. (6.2) and (6.3). In the forming process, wrap angle α changes with the increase of forming depth, and is determined by the geometrical parameters of the channel shape. According the geometrical relationship shown in Figure 6.9, the wrap angle is expressed as

$$\tan(\alpha) = \frac{h - t - (r_p + r_d)(1 - \cos(\alpha)) + t/\cos(\alpha)}{c + (r_p + r_d)(1 - \sin(\alpha))} \tag{6.4}$$

where t is the thickness of the metallic sheet.

6.3.2.1.2 Analysis of Strain and Thickness

Thickness, stress, and tension on the sheet vary during the stamping process. If an infinitesimal element is considered, the loading condition at an arbitrary point in the sheet metal is represented in Figure 6.12. As shown in Figure 6.12, direction 1 is along the sheet, direction 3 is in the thickness of the sheet, and direction 2 is perpendicular to the plane constituted by directions 1 and 3. During the manufacture of the channel, the length along direction 2 is very long and a plane-strain deformation ($\varepsilon_2 = 0$) is thus assumed.

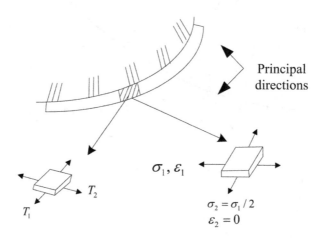

FIGURE 6.12
The main direction of an element of the sheet sliding on the face of the punch.

Defining $a' = \varepsilon_2/\varepsilon_1$ as the strain ratio, the effective strain is

$$\varepsilon_e = \sqrt{\frac{4}{3}\left(1 + b' + b'^2\right)}\varepsilon_1 = \frac{2}{\sqrt{3}}\varepsilon_1 \tag{6.5}$$

According to Levy-Mises flow rule [56], the relationship between the stress ratio a' ($a' = \sigma_2/\sigma_1$) and strain ratio b' is

$$a' = \frac{2b' + 1}{2 + b'} \tag{6.6}$$

From the earlier equation, the major principal tension T in the section plane is

$$T = \frac{2}{\sqrt{3}}t_0 k \sqrt{\left(\frac{2}{\sqrt{3}}\ln\left(\frac{t_0}{t}\right)\right)^n + I\eta^p} \exp\left(-\ln\left(\frac{t_0}{t}\right)\right) \tag{6.7}$$

Substitute Eq. (6.7) into Eqs. (6.2)–(6.4), the thickness distribution can be obtained. Figure 6.13 shows the schematic diagram of the thickness distribution. In the figure, it is found that part CD is the most dangerous point due to an excessive thinning caused by force concentration, and the crack will first occur at these points [57].

6.3.2.1.3 Volume Constancy Condition

In the forming of channels, points A and F are assumed to be fixed due to the geometric symmetry. The volume of the sheet metal between A and F is constant in the stamping process and thus the following is obtained:

$$V = V_0 - \int_A^F t\,dl = t_{AB}l_{AB} + t_{BC}l_{BC} + t_{CD}l_{CD} + t_{DE}l_{DE} + t_{EF}l_{EF} \tag{6.8}$$

where V_0 and V are the volume between A and F before and after stamping, respectively. $t_{AB}, t_{BC}, t_{CD}, t_{DE},$ and t_{EF} are the average thickness of $AB, BC, CD,$

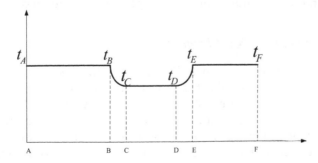

FIGURE 6.13
The schematic diagram of the thickness distribution.

DE, and EF, respectively. l_{AB}, l_{BC}, l_{CD}, l_{DE}, and l_{EF} are the length of AB, BC, CD, DE, and EF, respectively.

As shown in Figure 6.9, the process parameters of the die and punch are designed for efficiently fabricating the microchannel features. The length of each part can be formulated as follows:

$$
\begin{cases}
l_{AB} = \dfrac{w_p}{2} - r_p \\[4mm]
l_{BC} = \left(r_p + \dfrac{t_{BC}}{2} \right)\alpha \\[4mm]
l_{CD} = \sqrt{ \left(h - t_e - (r_p + r_d)(1-\cos\alpha) + \left(\dfrac{t_d}{2} + \dfrac{t_c}{2}\right)\cos\alpha \right)^2 + \left(c + (r_p + r_d)(1-\sin\alpha) - \left(\dfrac{t_d}{2} + \dfrac{t_c}{2}\right)\sin\alpha \right)^2 } \\[4mm]
l_{DE} = \left(r_d + \dfrac{t_{DE}}{2} \right)\alpha \\[4mm]
l_{EF} = \dfrac{w_s}{2} - r_d
\end{cases}
\tag{6.9}
$$

The volume before stamping is $V_0 = \dfrac{w_d + w_s}{2} t_0$.

Generally, the maximum channel height can be determined by the following steps: First, substituting Eq. (6.9) into Eq. (6.8), the relationship between the channel height h and the effective strain (direction 1) at point A, ε_A, is obtained, which could be used to calculate the principal tension at point A by Eq. (6.7). The principal tensions of T_C at point C and T_D at point D (most dangerous points) are obtained by Eqs. (6.2)–(6.4); Finally, ε_A and ε_D are calculated and compared with the instability criterion. Once the instability criterion is met, the maximum stamping height h_{max} is generated.

6.3.2.2 Design of BPP Flow Channels

In practice, the formability of metallic sheet is influenced by multiple factors, such as material property, sheet thickness, die design, and forming height. Figure 6.14 shows the section of the deformed flow channel in the stamping process. As shown in Figure 6.14, the formability of the metallic sheet is related to key geometric dimensions, which should be carefully considered and determined in die design. In this section, the groove width w_d and radius r_d of die, the radius r_p of punch, the clearance c between punch and die, and the width w_s of the rib are taken as design variables. The effect of

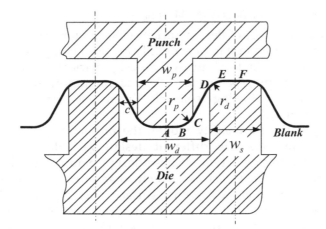

FIGURE 6.14
Schematic of the forming of metallic channels in stamping process.

the channel geometries on the reaction efficiency is also taken into account in the design of metallic BPPs.

The SS 304 with a thickness of 0.1 mm is widely used in the fabrication of BPPs. To determine the material mechanical property, the tensile sheet specimens with the gauge length of 20 mm and a free specimen length of 10 mm were cut from a flat cold-rolled sheet [58]. The uniaxial tensile experiments using the prepared specimens were conducted. The measured true stress–strain curves were determined and fitted by a power function as $\sigma_e = 1{,}532(\varepsilon_e)^{0.548}$.

On the other hand, the Taylor dislocation model involves an intrinsic material length l. According to prior research, burger vector b is given by about 0.3 nm for metals [59] and m is an empirical coefficient approximate of 0.6 [60]. By adopting these values of parameters, the intrinsic material length l is calculated as 5.57 μm. The mechanical properties, obtained by tensile experiment results, are shown in Table 6.1.

TABLE 6.1

Mechanical Properties of the SS304 Sheet

Mechanical Parameters	Value
Young's modulus E (GPa)	210
Poisson's ratio v	0.3
Initial yield stress σ_s (MPa)	361
Reference stress σ_{ref} (MPa)	1532
Exponent of power function n	0.548
Burgers vector b (nm)	0.3
Empirical coefficient m	0.6
Intrinsic material length l (μm)	5.57

6.3.2.2.1 Formability Analysis of the Metallic Sheet

To investigate the effect of design variables on the formability, the typical values commonly used in stamping of BPPs are used to represent their influence on formability. Table 6.2 gives the design variables with the consideration of real stamping parameters. The maximum forming height as the function of a few key tooling geometric dimensions is predicted based on the formability model.

According to the formability model, the prediction results are given to reflect the responses of different design variables and shown in Figure 6.15. Due to interaction of these design parameters, the response is considered by three aspects, in which some of the design parameters are fixed. Figure 6.15a is the response surface when $w_d = w_s = 1.0$ mm and $c = 0.2$ mm. It indicates that the maximum forming height increases with die radius r_d and punch radius r_p. Figure 6.15b is the response surface when $r_d = r_p = c = 0.2$ mm. The maximum forming height increases with the width w_d and the rib width w_s of the die. Figure 6.15c is the response surface when $r_d = r_p = 0.2$ mm and $w_d = 1.0$ mm, which illustrates that the maximum forming height is increased with rib width w and clearance c. The results shown in Figure 6.15 demonstrate that the decrease of these design parameters leads to a smaller channel height limit in the stamping process. If the channel height exceeds the forming limit, a rupture in the channel would occur.

6.3.2.2.2 Flow Channel Section Design

For the typical flow channel, as shown in Figure 6.5, the channel geometry could be described by the width w and the height h of the flow channel, and the width s and radius r of coolant channel, which affect the reaction efficiency of fuel cell and its manufacturability [61]. Therefore, the following geometric structure of flow channel is proposed by taking the reaction efficiency and the thin plate formability into consideration.

The formability model illustrates the relation between the channel forming limit and the stamping tools. Generally, as shown in Figure 6.14, the geometrical relationship between the desired channel and the stamping tools can be formulated as follows:

TABLE 6.2

Specifications of the Design Parameters (Unit: mm)

	Groove Width of Die w_d	Rib Width of Die w_s	Radius of Die r_d	Radius of Punch r_p	Clearance c
1	0.8	0.8	0.1	0.1	0.125
2	1.1	1.1	0.15	0.15	0.15
3	1.4	1.4	0.2	0.2	0.175
4	1.7	1.7	0.25	0.25	0.2
5	2.0	2.0	0.3	0.3	0.225

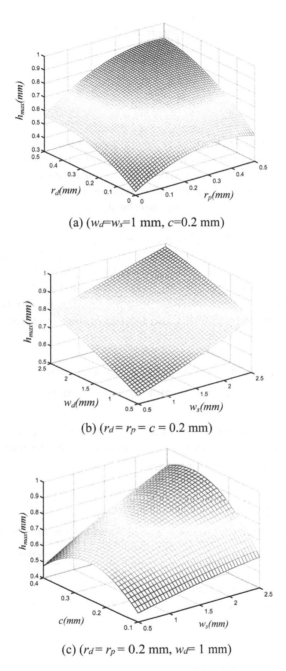

(a) $(w_d = w_s = 1 \text{ mm}, c = 0.2 \text{ mm})$

(b) $(r_d = r_p = c = 0.2 \text{ mm})$

(c) $(r_d = r_p = 0.2 \text{ mm}, w_d = 1 \text{ mm})$

FIGURE 6.15
Response surfaces of the maximum forming height.

$$\begin{cases} w_d = w + 2c \\ w_s = s \\ w_p = w \\ r_p = r_d = r \end{cases} \tag{6.10}$$

In Eq. (6.10), the general opening ratio $\varphi = w/(w + s)$ ranges from 45% to 75% to have a good tradeoff of the electrical contact resistance and the flow resistance of reaction gases. In this work, the opening ratio of 50% is adopted [62,63]. It means that the channel and rib have the same width. Considering the water transport and heat management, the size of $w = s = 0.9\,\text{mm}$ is adopted [62].

To ensure the production of channel height, the manufacturability of the channel is considered by the formability model. The clearance of the die is usually 1.5 times the thickness of the sheet ($c = 0.15\,\text{mm}$). Due to the high sensitivity of the manufacturability to radius, the radii of 0.1 mm and 0.15 mm are tested by the model. Submitting these parameters ($w = s = 0.9$, $c = 0.15\,\text{mm}$) and Eq. (6.10) into the formability model, the maximum forming limit is calculated as

$$\begin{cases} h_{\text{max}} = 0.54 \text{ mm} \quad \text{if} \quad r = 0.10 \text{ mm} \\ h_{\text{max}} = 0.54 \text{ mm} \quad \text{if} \quad r = 0.15 \text{ mm} \end{cases} \tag{6.11}$$

To ensure high production and reliability in the forming process, the radius is set as 0.15 mm and the channel height is 0.4 mm. The geometric parameters of BPP are obtained and presented in Table 6.3.

TABLE 6.3

The Geometric Parameters of BPP and the Die (unit: mm)

Parameter	Value
BPP	
Width of reaction channel w	0.9
Width of coolant channel s	0.9
Channel depth h	0.4
Radius r	0.15
Molds	
Groove width of die w_d	1.2
Rib width of die w_s	0.9
Radius of die r_d	0.15
Radius of punch r_p	0.15
Clearance c	0.15

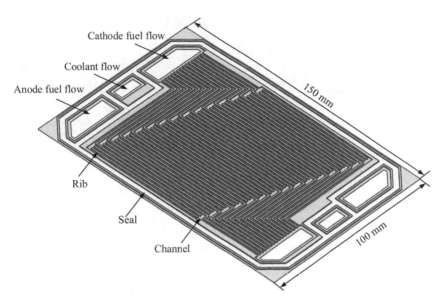

FIGURE 6.16
Schematics of the metallic BPP design.

According to the channel parameters in Table 6.3, the metallic BPP of 150 × 100 mm was designed and shown in Figure 6.16. A flow field with the parallel channels is constructed to distribute the reactant gas and coolant water uniformly on the cathode, anode, and cooling channels, respectively. The reaction area is 80 cm². The channel highlighted with yellow color represents the sealing groove.

6.3.2.3 Tooling Fabrication and BPP Manufacturing

6.3.2.3.1 Tooling Design and Fabrication

A set of stamping tooling, including punch and die, was designed and manufactured according to the BPP design, as shown in Figure 6.17. A hydraulic press machine with the capacity of 500 tons was used to fabricate the BPPs. A blank holder system between the punch and die was introduced to prevent the sheet wrinkle in the forming of geometries and features of the BPPs. Under deformation loading, the metallic sheet was first compressed by the blank holder system, and then the upper punch moved slowly toward the lower die, forcing the thin metallic sheet fill the microcavities on the die. Stripper springs were adopted to support the upper punch after the stamping to remove the formed BPPs from the die.

The SS 304 sheet with a thickness of 0.1 mm was used in the fabrication of BPPs. After microchannels were manufactured on the blank sheet using stamping process, the outside margin of the effective area was cut out by wire EDM. To provide outer channels for reactant gases and internal cooling

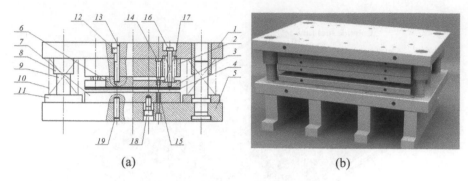

(a)   (b)

FIGURE 6.17
Assembly diagram of the stamping die: (a) the structure drawing of the die and (b) the assembled stamping die: 1. Upper die board; 2. Assistant plate; 3. Blank holding plate; 4. Blank holder; 5. The lower die plate; 6. Punch; 7. Die; 8. Guide bush; 9. Guide pillar; 10. Discharging spring; 11. Spring block; 12. Hexagon socket screw; 13. Pins; 14. Piercing punch; 15. Piercing die; 16. Bolt; 17. Blank holding spring; 18. Hexagon socket screw; 19. Pins.

required by fuel cell operation, the two pieces of the single plates were bonded together by fiber laser welding process. The final BPP was fabricated with the designed channel dimensions and the required specifications, as shown in Figure 6.18.

6.3.2.3.2 Forming Process and Analysis

Various forming forces of 600, 800 and 1,000 kN are employed in the fabrication of BPPs by forming process. In the process, a laser displacement sensor with a resolution of 0.2 μm in the vertical direction was used to measure the channel height. Figure 6.19 presents the profile curves of flow channel obtained by laser scanning under different forming forces, respectively. It is worth noting that the measured channels in one BPP have almost the same

FIGURE 6.18
Samples for stamped metallic BPPs.

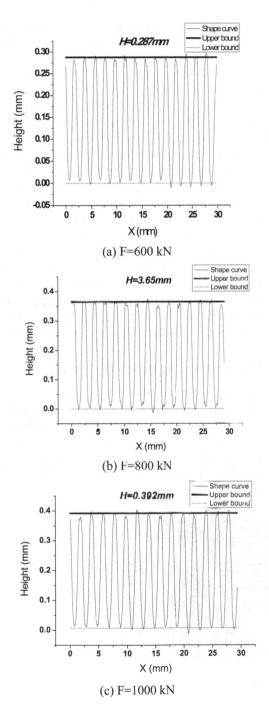

(a) F=600 kN

(b) F=800 kN

(c) F=1000 kN

FIGURE 6.19
The measurement results of the sectional topography.

height, which illustrates a high forming quality. The average channel height was calculated as 0.287, 0.365, and 0.392 mm, respectively. The higher channel height was obtained with a larger deformation load. Considering the designed channel height of 0.4 mm, the deformation loading of 1,000 kN was selected to ensure the desired height of the channel.

The thickness distribution at some key points was determined to analyze the quality of the formed BPPs. The specimens of the half flow channel were obtained from the formed workpieces by using wire EDM. Grinding and polishing were used for machining the surfaces of BPPs before measurement. The thicknesses at the five typical points, viz., points A, B, C, D, and E, as shown in Figure 6.20a, were measured and compared with the predicted results by the formability model. As shown in Figure 6.20b, the predicted results show a good agreement with the experimental ones. The thicknesses at point A (0.0934 mm), point C (0.0908 mm), and point E (0.0941 mm) have a little difference from the initial blank sheet (0.10 mm). The thinnest region occurred at the place where the sheet and die fillet contact is points B and D with the thickness of 0.0820 and 0.0798 mm. The minimum thickness and its location are consistent with the design principles proposed earlier. In addition, the largest blank thinning was about 20% of the initial thickness, which is safe for production. Therefore, the numerical model proposed is feasible to guide the fabrication of BPPs.

6.3.3 Meso- and Microstamping with Soft Punch

Compared with the traditional sheet forming processes with rigid die and punch, stamping with soft punch only employs one rigid die and the punch is made of soft materials such as natural or synthetic rubbers [54]. As shown in Figure 6.21 [43], only the rigid die needs to be designed and accurately manufactured. In addition, the soft punch and rigid die can be precisely assembled. As a result, the manufacturing lead time and production cost can be greatly reduced.

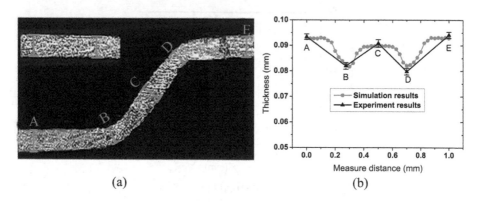

(a)   (b)

FIGURE 6.20
Thickness distribution of the microchannel: (a) key measured points and (b) comparison between the deformed results and the predicted ones by the model.

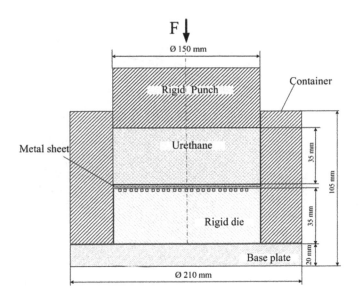

FIGURE 6.21
Sketch of the meso- and microstamping with soft punch [43].

In this section, a novel configuration design of metallic BPPs with an inter-digitated flow field was first proposed and shown in Figure 6.22 [43]. The configuration design has been optimized to provide a uniform gas flow pressure drop and an effective water management in the fuel cell operation. Another problem is the feasibility of fabrication of this design by soft stamping process, especially in the critical region that is prone to have fracture due to stress concentration and uneven distribution of strain in the deformation process. Generally, die design is the critical activity that determines whether the high-quality workpiece can be obtained or not. Hence, FE simulation was used to investigate the formability of BPPs.

6.3.3.1 Flow Channel Design for Manufacturability

6.3.3.1.1 Finite Element Analysis

The BPP contains many microchannels, where the fracture could occur at the corner of ribs due to the large stress concentration and uneven deformation in the forming process, as shown in Figure 6.22 [43]. Therefore, the rigid die used in soft stamping to fabricate the BPP should be carefully designed, especially for the contact region with the heads of BPPs' ribs presented in Figure 6.23a [43].

Along the section line as shown in Figure 6.23a [43], the cutaway view can be obtained and the key geometric dimensions should be considered in the die design. Figure 6.23b shows the key dimensions of the section of the rigid die used to form the BPPs [43]. The key dimensions include the main

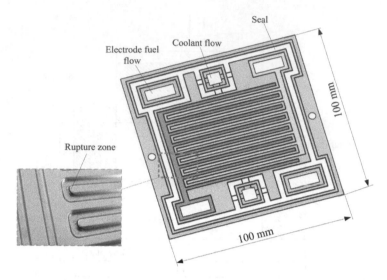

FIGURE 6.22
Sketch of the thin metallic BPP's with the interdigitated flow field and fracture in forming process [43].

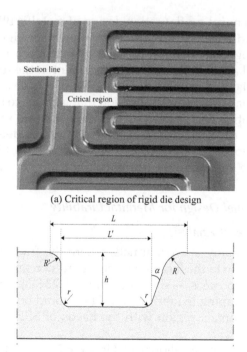

(a) Critical region of rigid die design

(b) Sketch of the section view

FIGURE 6.23
Key dimensions in the cutaway view of the rigid die along the section line [43].

channel width L' and L, channel depth h, upper transition radius R and R', lower radius r and draw angle a. Some of these dimensions should be maintained according to the design requirements for high compactness, reaction performance, and easy stack assembly, as shown in Table 6.4. Other dimensions dominating the overall formability, represented by draw angle a and upper transition radius R, should be investigated carefully to reduce the risk of material fracture.

6.3.3.1.2 Channel Section Design

According to the key dimensions and materials mentioned earlier, the parameterized FE 3D models were established, in which the upper transition radius and the draw angle were parameterized under four different cases listed in Table 6.5, to evaluate its formability. Forming process simulations were conducted to be compared with the actual forming experiments.

The sheet thickness of the formed workpiece was calculated to evaluate the fracture risk. The minimum thickness in the dangerous zone, where the force concentrate and rupture occurs, was obtained and shown in Table 6.5. According to the thinning rates and the material properties, the risk of fracture can be evaluated. The minimum thickness for the four cases is 0.643, 0.712, 0.788, and 0.815 mm, respectively. It indicates that the thinning of the formed parts reduced with the increase of the upper transition radius and the draw angle.

To validate the accuracy of the simulation results, Cases 1 and 3 were chosen for the real sheet forming experiments. The die setups were designed and fabricated according to the geometric dimensions described earlier. The experiments were conducted, where the formed parts were obtained. The workpiece was cut along the section to analyze its formability. Figure 6.24 shows the experimental and numerical results of the workpiece for Case 1 [43].

TABLE 6.4

Key Geometric Dimensions for the Soft Punch Stamping Process

Parameters	L'	L	r	R'	R'	h
Value (mm)	1.6	3.05	0.15	0.25	0.4	0.5

TABLE 6.5

Key Geometric Dimensions and Formability Evaluation

	Key Dimensions of the Die		The Formed Workpiece		
Case	Upper Transition Radius R (mm)	Draw Angle a (°)	Minimum Thickness (mm)	Thinning (%)	Rupture Evaluation
1	0.2	0	0.643	35.7	Rupture
2	0.4	0	0.712	29.8	Dangerous
3	0.2	30	0.788	21.2	Acceptable
4	0.4	30	0.815	18.5	Safe

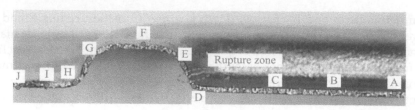

(a) Formed workpiece

(b) Numerical simulation results

FIGURE 6.24
Experimental and numerical results for the workpiece (Case 1) [43].

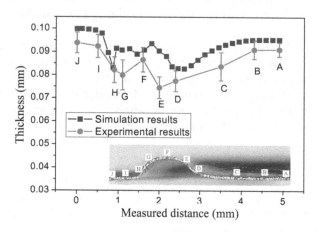

FIGURE 6.25
Thickness distribution of the formed parts (Case 3) [43].

From the experimental results, it was found that the fracture occurs at the rupture zone consistent with the simulation results. By the same method, the workpiece for Case 3 could be achieved, and its thickness was measured by a high-resolution Flexivision measurement system. Figure 6.25 illustrates the comparison of the experimental and simulation results [43].

From the analysis presented earlier, it is found that the large upper transition radius and draw angle in the die design could reduce force concentration. As a result, the formability of the workpeice could be improved.

6.3.3.2 Formability of BPPs by Soft Punch Stamping

Generally, the accurate numerical simulations should be performed before physical experiments to support design in such a way to save manufacturing cost and cut development lead time. Based on the simulation results, some defects can be predicted, and the process parameter modification can be done to improve the quality of the final workpiece.

According to the design guidelines for the die critical region discussed earlier and the configurations of BPPs, 3D numerical simulation models were established based on the commercial FE code LS-DYNA. The geometric dimensions of the rigid die in the critical region are consistent with Case 4, which shows high formability. The same as the sketch diagram shown in Figure 6.21 [43], the diameters of the rigid die, the soft material, the initial blank, and the inner diameter of the container were all 150 mm. The material models of the soft tool and the sheet metal were developed in accordance with the material properties used in the physical experiments. The soft tool is urethane rubber with 70 Shore A hardness and the metal sheets is SS304 with a thickness of 0.1 mm. Half of the physical model has been analyzed because of the symmetric structure design of the BPP, and the die was defined as a rigid body to simplify the numerical simulation model.

The forming limit diagram and the thickness distribution were used to evaluate the feasibility of fabricating metallic BPPs by soft punch stamping. From the forming limit diagram, it is found that most of the combinations of major and minor strains in the effective area (100 mm × 100 mm) at the end of forming fall into the safe zone due to the proper die design, material selection, and process planning, which means that there is no wrinkle and fracture in the sheet metal, as shown in Figure 6.26a [43]. On the other hand, the thickness distribution also decides whether a crack will occur or not with the increasing thinning of the metal sheets during the stamping process. Figure 6.26b [43] shows the final thickness distribution at the end of the forming process and the minimum thickness is 0.081 mm and the maximum thinning is 19%. These results indicate that the process is safe and acceptable.

6.3.3.3 BPP Manufacturing and Analysis

According to the design principles obtained by numerical simulations mentioned earlier, the rigid die was fabricated, and the experimental setups were carefully prepared. The SS sheet of SS304 with a thickness of 0.1 mm was used in experiment. The soft punch used in the experiments is 70 Shore A hardness urethane rubber. The forming experiments were conducted by using a 500-ton hydraulic press machine.

The single plate can be deformed by using the meso- and microstamping with soft punch. Two symmetric plates were joined together by welding, and the outside margin of the effective area was cut out by wire EDM. The inlets

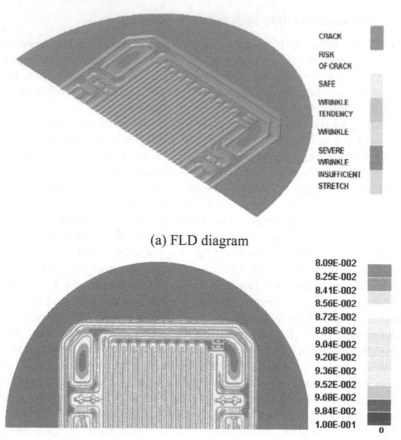

(a) FLD diagram

(b) Thickness distribution of the half formed workpiece

FIGURE 6.26
Formability analysis of BPPs by numerical simulation [43].

and outlets were excavated according to the designed dimensions. With the efficacious surface-coating process, the high-quality BPPs, as shown in Figure 6.27 [64], were fabricated eventually.

The correctness of the design guidelines were thus validated by experiments. The critical region, where the force is concentrated and the fracture usually happens, was extracted by wire EDM. Figure 6.14 shows the comparison of the experimental workpiece and the numerical summation results. Moreover, the thickness of the section cut along the section line was measured in detail at the key points marked in Figure 6.28 [43]. The experimental results show a good agreement with the numerical simulation results and are shown in Figure 6.29 [43]. The minimum thickness and its location are also in accordance with the design principles proposed earlier.

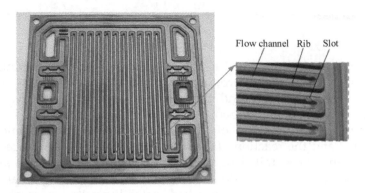

FIGURE 6.27
The formed parts by using the soft punch stamping process [64].

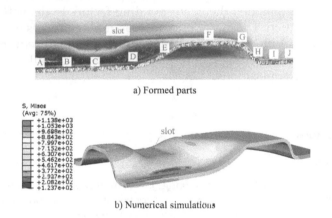

a) Formed parts

b) Numerical simulations

FIGURE 6.28
Section analysis for slotted-interdigitated shape [43].

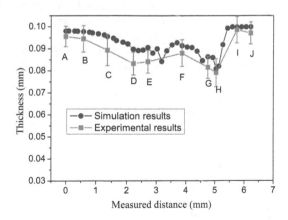

FIGURE 6.29
Thickness of the section for BPPs [43].

6.4 Summary

In this chapter, the concerned issues in meso- and microstamping of the miniaturized sheet metal products for industrial applications are presented. As a typical meso- and microstamped product, a complete design and fabrication of the stamped metallic BPPs in the fuel cell were conducted with a consideration of the formability of meso- and microscale sheet metal and the reaction efficiency of the fuel cell. The process characteristics and the detailed deformation behaviors of the rigid and soft stamping processes were systematically described. By using an analytical formability model with the prescribed instability criterion, strain, thickness distribution, and limited channel design, punch and die fabrication of the formed workpiece in rigid stamping process were analyzed. Based on a developed FE model, the corresponding factors in the soft stamping process were studied. The meso- and microstamping process has a promising potential for wide applications, while the functional requirements of products and microforming-technique-related issues should be seriously considered in its applications.

References

1. L.F. Peng, X.M. Lai, H.J. Lee, J.H. Song, J. Ni, Analysis of micro/mesoscale sheet forming process with uniform size dependent material constitutive model, *Materials science and Engineering: A* 526(1) (2009) 93–99.
2. W. Chan, M.W. Fu, B. Yang, Study of size effect in micro-extrusion process of pure copper, *Materials & Design* 32(7) (2011) 3772–3782.
3. M.W. Fu, B. Yang, W.L. Chan, Experimental and simulation studies of micro blanking and deep drawing compound process using copper sheet, *Journal of Materials Processing Technology* 213(1) (2013) 101–110.
4. M. Thome, G. Hirt, B. Rattay, Metal flow and die filling in coining of micro structures with and without flash, *Advanced Materials Research* 6–8 (2005) 631–638.
5. A.R. Razali, Y. Qin, A review on micro-manufacturing, micro-forming and their key issues, *Procedia Engineering* 53 (2013) 665–672.
6. F. Gong, B. Guo, C.J. Wang, D.B. Shan, Micro deep drawing of micro cups by using DLC film coated blank holders and dies, *Diamond and Related Materials* 20(2) (2011) 196–200.
7. J.G. Liu, M.W. Fu, J. Lu, W.L. Chan, Influence of size effect on the springback of sheet metal foils in micro-bending, *Computational Materials Science* 50(9) (2011) 2604–2614.
8. X. Peng, Y. Qin, R. Balendra, Analysis of laser-heating methods for micro-parts stamping applications, *Journal of Materials Processing Technology* 150(1–2) 2004, 84–91.

9. U.M. Attia, J.R. Alcock, Fabrication of hollow, 3D, micro-scale metallic structures by micro-powder injection moulding, *Journal of Materials Processing Technology* 212(10) (2012) 2148–2153.

10. J. Cao, N. Krishnan, Z. Wang, H.S. Lu, W.K. Liu, A. Swanson, Microforming: Experimental investigation of the extrusion process for micropins and its numerical simulation using RKEM, *Journal of Manufacturing Science and Engineering* 126(4) (2004) 642–652.

11. U. Engel, R. Eckstein, Microforming—From basic research to its realization, *Journal of Materials Processing Technology* 125–126(Supplement C) (2002) 35–44.

12. A. Tonkovich, D. Kuhlmann, A. Rogers, J. McDaniel, S. Fitzgerald, R. Arora, T. Yuschak, Microchannel technology scale-up to commercial capacity, *Chemical Engineering Research and Design* 83(6) (2005) 634–639.

13. P. Reuse, A. Renken, K. Haas-Santo, Görke O., K. Schubert, Hydrogen production for fuel cell application in an autothermal micro-channel reactor, *Chemical Engineering Journal* 101(1) (2004) 133–141.

14. LIPSKI R, Smaller-scale GTL enters the mainstream, (2017) http://www.gas-processingnews.com/.

15. S.P. Jung, C.L. Lee, C.C. Chen, W.S. Chang, C.C. Yang, Development of novel proton exchange membrane fuel cells using stamped metallic bipolar plates, *Journal of Power Sources* 283 (2015) 429–442.

16. W. Ehrfeld, V. Hessel, S. Kiesewalter, H. Löwe, Th Richter, J. Schiewe, *Implementation of Microreaction Technology in Process Engineering, Microreaction Technology: Industrial Prospects: IMRET 3: Proceedings of the Third International Conference on Microreaction Technology*, W. Ehrfeld, ed., Springer Berlin Heidelberg, Berlin, Heidelberg, 2000, pp. 14–34.

17. K.K. Yeong, A. Gavriilidis, R. Zapf, V. Hessel, Catalyst preparation and deactivation issues for nitrobenzene hydrogenation in a microstructured falling film reactor, *Catalysis Today* 81(4) (2003) 641–651.

18. K. Jähnisch, M. Baerns, V. Hessel, W. Ehrfeld, V. Haverkamp, H. Löwe, Ch Wille, A. Guber, Direct fluorination of toluene using elemental fluorine in gas/liquid microreactors, *Journal of Fluorine Chemistry* 105(1) (2000) 117–128.

19. D.K. Qiu, P.Y. Yi, L.F. Peng, X.M. Lai, Channel dimensional error effect of stamped bipolar plates on the characteristics of gas diffusion layer contact pressure for proton exchange membrane fuel cell stacks, *ASME Journal of Fuel Cell Science and Technology* 12(4) (2015) 041002.

20. S. Mahabunphachai, Ö.N. Cora, M. Koc, Effect of manufacturing processes on formability and surface topography of proton exchange membrane fuel cell metallic bipolar plates, *Journal of Power Sources* 195(16) (2010) 5269–5277.

21. D.K. Qiu, P.Y. Yi, L.F. Peng, X.M. Lai, Study on shape error effect of metallic bipolar plate on the GDL contact pressure distribution in proton exchange membrane fuel cell, *International Journal of Hydrogen Energy* 38(16) (2013) 6762–6772.

22. Q.H. Hu, D.M. Zhang, H. Fu, K.K. Huang, Investigation of stamping process of metallic bipolar plates in PEM fuel cell-Numerical simulation and experiments, *International Journal of Hydrogen Energy* 39 (2014) 13770–13776.

23. F. Vollertsen, Z. Hu, H.S. Niehoff, C. Theiler, State of the art in micro forming and investigations into micro deep drawing, *Journal of Materials Processing Technology* 151(1) (2004) 70–79.

24. K. Shaji, S.K. Das, Effect of plate characteristics on axial dispersion and heat transfer in plate heat exchangers, *Journal of Heat Transfer* 135(4) (2013) 041801.

25. C.H. Ahn, J. Choi, C. Son, J.K. Min, S.H. Park, D. Gillespie, J.S. Go, Measurement of pressure distribution inside a cross-corrugated heat exchanger using micro-channel pressure tappings, *Measurement Science and Technology* 24(3) (2013) 035306.
26. N. Gallandat, D. Hesse, M.J. Rhett, Microfeature heat exchanger using variable-density arrays for near-isothermal cold-plate operation, *Journal of Electronic Packaging* 138(1) (2016) 010908.
27. K.W. Roth, D. Westphalen, J. Dieckmann, S.D. Hamilton, W. Goetzler, *Energy Consumption Characteristics of Commercial Building HVAC Systems Volume III: Energy Savings Potential*, US Department of Energy, 2002.
28. A.L. Tonkovich, G. Roberts, S.P. Fitzgerald, T.M. Werner, M.B. Schmidt, R.J. Luzenski, G.B. Chadwell, J.A. Mathias, A. Gupta, D.J. Kuhlmann, Microchannel apparatus, methods of making microchannel apparatus, and processes of conducting unit operations, Google Patents, 2006.
29. H.C. Cho, B.J. Na, C.H. Ahn, B.S. Shin, S.H. Park, C.M. Son, H.S. Moon, J.S. Go, Development of measurement method of pressure distribution inside a compact and complex heat exchanger, *Flow Measurement and Instrumentation* 46(Part A) 2015, 93–102.
30. O.K. Siddiqui, S.M. Zubair, Efficient energy utilization through proper design of microchannel heat exchanger manifolds: A comprehensive review, *Renewable and Sustainable Energy Reviews* 74(Supplement C) 2017, 969–1002.
31. B. Alm, R. Knitter, J. Haußelt, Development of a ceramic micro heat exchanger – Design, construction, and testing, *Chemical Engineering & Technology* 28(12) (2005) 1554–1560.
32. M.R.H. Abdel-Salam, R.W. Besant, C.J. Simonson, Sensitivity of the performance of a flat-plate liquid-to-air membrane energy exchanger (LAMEE) to the air and solution channel widths and flow maldistribution, *International Journal of Heat and Mass Transfer* 84(Supplement C) 2015, 1082–1100.
33. W. Ehrfeld, V. Hessel, V. Haverkamp, *Microreactors*, Wiley Online Library, 2000.
34. V. Hessel, H. Löwe, Microchemical engineering: Components, plant concepts, user acceptance–Part III, *Chemical Engineering & Technology* 26(5) (2003) 531–544.
35. P. Löb, K.S. Drese, V. Hessel, S. Hardt, C. Hofmann, H. Löwe, R. Schenk, F. Schönfeld, B. Werner, Steering of liquid mixing speed in interdigital micro mixers—From very fast to deliberately slow mixing, *Chemical Engineering & Technology* 27(3) (2004) 340–345.
36. W.K. Tseng, J.L. Lin, W.C. Sung, S.H. Chen, G.B. Lee, Active micro-mixers using surface acoustic waves on Y-cut 128° LiNbO$_3$, *Journal of Micromechanics and Microengineering* 16(3) (2006) 539.
37. N.T. Nguyen, Z.G. Wu, Micromixers—A review, *Journal of Micromechanics and Microengineering* 15(2) (2005) R1.
38. S. Lee, H. Jeong, B.K. Ahn, T.W. Lim, Y.J. Son, Parametric study of the channel design at the bipolar plate in PEMFC performances, *International Journal of Hydrogen Energy* 33(20) (2008) 5691–5696.
39. Metallic BPP by Borit Inc., (2016) https://www.borit.be/.
40. Metallic BPP by SJTU- Shanghai Zhizhen Inc., (2017) http://shanghaizhizhen.com/.
41. Metallic BPP by Tallistech Inc., (2017) http://www.tallistech.com/.

42. M.F. Peker, Ö.N. Cora, M. Koc, Investigations on the variation of corrosion and contact resistance characteristics of metallic bipolar plates manufactured under long-run conditions, *International Journal of Hydrogen Energy* 36(23) (2011) 15427–15436.

43. L.F. Peng, X.M. Lai, P.Y. Yi, J.M. Mai, J. Ni, Design, optimization, and fabrication of slotted-interdigitated thin metallic bipolar plates for PEM fuel cells, *ASME Journal of Fuel Cell Science and Technology* 8(1) (2011) 011002.

44. L.F. Peng, X.M. Lai, D.A. Liu, P. Hu, J. Ni, Flow channel shape optimum design for hydroformed metal bipolar plate in PEM fuel cell, *Journal of Power Sources* 178(1) (2008) 223–230.

45. S.J. Lee, C.Y. Lee, K.T. Yang, F.H. Kuan, P.H. Lai, Simulation and fabrication of micro-scaled flow channels for metallic bipolar plates by the electrochemical micro-machining process, *Journal of Power Sources* 185(2) (2008) 1115–1121.

46. J.C. Hung, T.C. Yang, K.C. Li, Studies on the fabrication of metallic bipolar plates—Using micro electrical discharge machining milling, *Journal of Power Sources* 196(4) (2011) 2070–2074.

47. P. Zhang, M. Pereira, B. Rolfe, W. Daniel, M. Weiss, Deformation in micro roll forming of bipolar plate, *Journal of Physics: Conference Series* 896(1) (2017) 012115.

48. J. Shang, L. Wilkerson, S. Hatkevich, G.S. Daehn, Commercialization of fuel cell bipolar plate manufacturing by electromagnetic forming, 2010.

49. L.F. Peng, P.Y. Yi, X.M. Lai, Design and manufacturing of stainless steel bipolar plates for proton exchange membrane fuel cells, *International Journal of Hydrogen Energy* 39(36) (2014) 21127–21153.

50. K. Pallav, P. Han, J. Ramkumar, Nagahanumaiah, K.F. Ehmann, Comparative assessment of the laser induced plasma micromachining and the micro-edm processes, *Journal of Manufacturing Science and Engineering* 136(1) (2013) 011001.

51. V.K. Jain, S. Kalia, A. Sidpara, V.N. Kulkarni, Fabrication of micro-features and micro-tools using electrochemical micromachining, *International Journal of Advanced Manufacturing Technology* 61(9–12) (2012) 1–9.

52. A.P. Manso, F.F. Marzo, J. Barranco, X. Garikano, M.M. Garmendia, Influence of geometric parameters of the flow fields on the performance of a PEM fuel cell. A review, *International Journal of Hydrogen Energy* 37(20) (2012) 15256–15287.

53. R. Anderson, L.F. Zhang, Y.L. Ding, M. Blanco, X.T. Bi, D.P. Wilkinson, A critical review of two-phase flow in gas flow channels of proton exchange membrane fuel cells, *Journal of Power Sources* 195(15) (2010) 4531–4553.

54. L.F. Peng, D.A. Liu, P. Hu, X.M. Lai and J. Ni, Fabrication of metallic bipolar plates for proton exchange membrane fuel cell by flexible forming process-numerical simulations and experiments, *ASME Journal of Fuel Cell Science and Technology* 7(3) (2010) 031009.

55. M. Koc, S. Mahabunphachai, Feasibility investigations on a novel micro-manufacturing process for fabrication of fuel cell bipolar plates: Internal pressure-assisted embossing of micro-channels with in-die mechanical bonding, *Journal of Power Sources* 172(2) (2007) 725–733.

56. H. Gao, Y. Huang, W.D. Nix, J.W. Hutchinson, Mechanism-based strain gradient plasticity—I. Theory, *Journal of the Mechanics and Physics of Solids* 47(6) (1999) 1239–1263.

57. L.F. Peng, P.Y. Yi, P. Hu, X.M. Lai, J. Ni, Analysis of micro/mesoscale sheet forming process by strain gradient plasticity and its characterization of tool feature size effects, *Journal of Micro and Nano-Manufacturing* 3(1) (2015) 011006.

58. L.F. Peng, P. Hu, X.M. Lai, D.Q. Mei, J. Ni, Investigation of micro/meso sheet soft punch stamping process—Simulation and experiments, *Materials & Design* 30(3) (2009) 783–790.

59. S.P. Lele, L. Anand, A large-deformation strain-gradient theory for isotropic viscoplastic materials, *International Journal of Plasticity* 25(3) (2009) 420–453.

60. Y. Huang, Z. Xue, H. Gao, W.D. Nix, Z.C. Xia, A study of microindentation hardness tests by mechanism-based strain gradient plasticity, *Journal of Materials Research* 15(8) (2000) 1786–1796.

61. L.F. Peng, X.M. Lai, D.A. Liu, P. Hu, J. Ni, Flow channel shape optimum design for hydroformed metal bipolar plate in PEM fuel cell, *Journal of Power Sources* 178(1) (2008) 223–230.

62. S. Shimpalee, Z.J.W. Van, Numerical studies on rib & channel dimension of flow-field on PEMFC performance, *International Journal of Hydrogen Energy* 32(7) (2007) 842–856.

63. D.K. Qiu, L.F. Peng, P.Y. Yi, X.M. Lai, A micro contact model for electrical contact resistance prediction between roughness surface and carbon fiber paper, *International Journal of Mechanical Sciences* 124–125 (2017) 37–47.

64. L.F. Peng, J.M. Mai, P. Hu, X.M. Lai, Z.Q. Lin, Optimum design of the slotted-interdigitated channels flow field for proton exchange membrane fuel cells with consideration of the gas diffusion layer intrusion, *Renewable Energy* 36(5) (2011) 1413–1420.

7

Meso- and Microscaled Bulk Components Produced by Progressive and Compound Forming of Sheet Metals

7.1 Introduction

With the ever-increasing demand on integral design and development of complex structures and components and the bullish trends of ubiquitous product miniaturization, energy saving, and weight reduction in different industrial clusters, the integral design and development of micro- and mesoscaled parts and components with complex geometries and features for different application sectors, including electronics, automobiles, biomedicine, aerospace, and energy industries [1,2], have become crucial, as design and development of these types of miniaturized parts and components need to address the issues arising from integration of function, property, and geometry of the developed parts and manufacturing issues coming from the downsize of geometry scale and variation of microstructure of materials. To fabricate miniaturized parts, micromanufacturing technologies including micromachining [3], microinjection molding [4,5], powder injection molding [6–8], and microscaled plastic deformation, i.e., microforming [1,2,9–11], have been successfully developed. Among these production approaches, meso- and microforming, which fabricate meso- and microscaled parts with the desirable geometries via plastic deformation, have attracted increasing attention due to the excellent mechanical properties of the fabricated parts and the low production cost and high productivity. Therefore, a large-scale mass production of meso- and microscaled parts via meso- and microforming has become a critical and promising approach.

For using meso- and microforming to produce downsized bulk parts, there are a few critical issues. If these issues cannot be efficiently addressed, the promising potential of meso- and microforming would be crippled. The first one is the difficulty to control the accuracy of geometry and shape of meso- and microformed parts due to the anisotropies of material properties and deformation behaviors. From a deformation perspective, it is well known that single grain deforms via slipping. For a given stress direction, there is a preferred slip plane and direction, which make the deformation anisotropic. In macroscale deformation, the materials are composed of a large number of grains. From a statistic point of view, different grains have different sizes, orientations, and forms and can be evenly distributed inside the materials. In a microscale one, however, the forming material may have only a few grains, and the properties and behaviors of individual grains could be different in different directions, which lead to inhomogeneous deformation and irregular behaviors in deformation, and thus lead to formation of irregular geometry and shape of the deformed parts. Upon completion of forming, the springback caused by the elastic recovery of the deformed meso- and microscaled also affect the dimensional accuracy, which is more complicated than that in macroscale. In meso- and microscaled forming, the relationship of elastic recovery with geometry size and scale factor in foil bending is shown in Figure 7.1. From Figure 7.1a, it is found that the elastic recovery is decreased with foil thickness, but it increases when the thickness is less than 100 μm. In Figure 7.1b, it shows the relationship of elastic recovery with scaling factor λ, which has the same variation trend as that in Figure 7.1a. In addition, the materials with fine and coarse grains have the same trend. This complicated elastic recovery makes the prediction and determination of accurate geometries and dimensions in design stage more difficult.

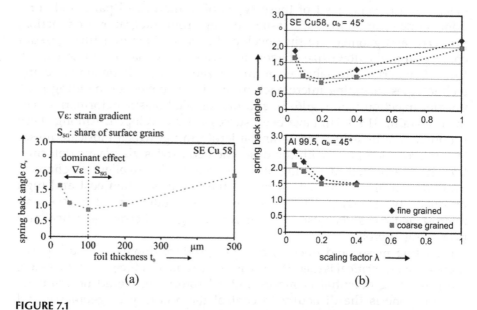

FIGURE 7.1
The recovery with geometry size and size factors: (a) the relationship between springback and foil thickness [12] and (b) the relationship between springback and scaling factor [13].

The second issue is the preparation of billet for meso- and microforming. Since the billet is also in meso- and microscale due to the constant material volume in deformation, preparation of small-scaled billet or preform is also a critical issue. Traditionally, the main approach is micromachining to prepare the microbillet, but it is time consuming and has low productivity. In addition, there are two more alternative approaches. The first one is to cut billets from the metal wire directly via simple shearing [14], while the second one is to punch sheet metal directly to form billets [15]. For the first one, however, it is difficult to control the quality of shearing surface via shear fracture and the inclined fracture surface, which will be the top surface of the billet, as it would generate the eccentric force in the interaction between punch and billet in the subsequent forming process, which could break the micropunch in the deformation process. For the second approach, it is more promising. But the dimensions controlled by the piercing hole of die are more accurate than that in the punching direction. Due to the dimension scattering or fluctuation in the punching direction, the volume produced by this approach is not constant and could be difficult to control. In addition, individual billets are placed in the die cavity for meso- or microforming, and the positioning along the right place and alignment in the vertical direction is a nontrivial issue. Since the weight of billet is small, the billets may stick onto the surface of the handling tool or the wall of cavity and could be difficult to place in the right position. So is the alignment. All of these need to be addressed successfully before this approach to produce billet becomes promising and has a good potential to be used in industries.

Another crucial issue is ejection of the microformed parts from die cavity. Upon completion of microforming using the individual billets, the ejection of the deformed microparts from the die cavity is very difficult, if not impossible, due to the elastic recovery, especially for extrusion-based deformation in the process, and the friction and interaction between billet surface and die cavity. The often encountered problems are warpage and distortion of microparts, part surface damage and scrape, die severe wearing, damage of ejection system, etc. caused by ejection action. For mass production, all of these would easily cripple or disable the whole forming system. Therefore, seeking for an efficient meso- and microforming method for mass production of meso- and microscaled bulk parts using plastic deformation is important and critical in the metal-forming arena.

In this chapter, a promising progressive and compound meso- and microforming process is succinctly introduced, and its details in terms of process, tooling, deformation, and product quality are systematically described and presented. The progress is first progressive in nature, which means the deformation is realized step by step from a horizontal dimension perspective, while it is compound deformation by which one vertical operation realizes a few deformations such as extrusion and blanking simultaneously from a vertical dimension aspect. The main uniqueness is the bulk forming of sheet metals to make the volume-based meso- and microscaled parts. The key characteristics include (a) avoiding the preparation of billets for meso- and microforming; (b) facilitating the transportation of billet or preform between different deformation stations; (c) easy positioning, placing, and alignment of preform in subsequent operations; (d) no ejection issue articulated above and thus less damage on the surface and dimension accuracy of the deformed parts; (e) improving productivity of the process; and (f) prolonging die service life, etc.

However, there are also some disadvantages for this developed meso- and microscaled forming processes. The most critical one is the dimensional accuracy of the deformed parts. In the meso- and microscaled progressive and compound forming system, the horizontal dimensions are controlled by the surfaces of punch, die cavity, and die insert, and thus the dimensions of the deformed parts are quite stable. The vertical dimensions, however, are basically not restricted by die features, and therefore, the material flowing in the vertical direction have more freedom and the amount of material flowing downwards is affected by many factors including microstructural grain size, deformation pattern, tooling design and geometries, lubrication between deforming material and tooling, and forming process parameters. In this chapter, how these factors affect the dimensions and geometries of deformed parts will also be analyzed.

7.2 Progressive and Compound Bulk Forming of Sheet Metals

The progressive and compound bulk forming of sheet metals is a unique forming process to produce bulk components by directly using sheet metals. The uniqueness lies in the forming implemented in two dimensions, viz., horizontal and vertical dimensions, in the forming system. The former realizes the station-based deformation in different operation stops progressively. In this way, the feeding of sheet metal is realized automatically. In the design of different deformations in each individual deformation station, the formability and deformation behaviors of materials in the designed deformation allocation and die geometries and structure need to be considered. For the deformation in vertical dimension, the compound deformation is designed to realize different operations such as the combination of piercing for positioning and extrusion for deformation, or extrusion for deformation and blanking for trimming the deformed part away from the sheet metal. Bulk forming of sheet metals is in such a progressive and compound way to develop the aforementioned characteristics with the embedded disadvantages.

7.2.1 Meso- and Microscaled Bulk Components

In this book, mesoscaled bulk components are defined as the deformed parts with at least two geometrical dimensions falling in between 1 and 10 mm; while the parts with the dimensions exceeding this requirement are termed as macroscaled components. In terms of microscaled parts, they have at least two dimensions less than 1 mm. In progress and compound-formed bulk parts with meso- and microscales, they can be classified into hollow and solid parts with different size scales as shown in Figure 7.2a, b. In addition, from feature, shape and geometry perspective, the bulk parts can be further classified into different categories such as single, double, and multiflanged bulk parts with different size scales in Figure 7.2c–e. Furthermore, from vertical deformation aspect, the bulk parts can be made with internal profile as shown in Figure 7.2f (top view of the part) and the parts with both internal and external profiles in Figure 7.2g. In forming process determination, deformation allocation, punch and die design, simultaneous consideration of process route and operation station design, allocation of deformation in each station and its realization, and design of punch and die feature, geometry and shape should be done in upfront stage in the entire design of the progressive and compound forming system.

7.2.2 Progressive and Compound Meso- and Microforming System

Figure 7.3 shows the progressive and compound sheet metal forming for making hollow and solid bulk parts directly using sheet metal. Figure 7.3a shows the process for making hollow-type bulk parts. The first operation shown in Figure 7.3a is the shearing operation proposed for making hollow

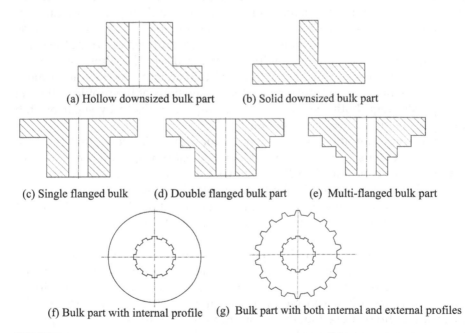

(a) Hollow downsized bulk part (b) Solid downsized bulk part

(c) Single flanged bulk (d) Double flanged bulk part (e) Multi-flanged bulk part

(f) Bulk part with internal profile (g) Bulk part with both internal and external profiles

FIGURE 7.2
Bulk parts formed by progressive and compound forming of sheet metals.

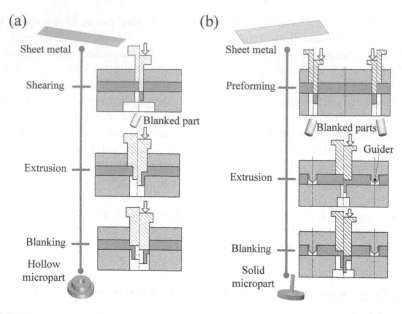

FIGURE 7.3
Shearing process customized for progressive forming: (a) hollow bulk part and (b) solid bulk part [16].

profile with different operations in each station, which includes shearing for making hole profiles by piercing or producing blanked part via blanking. In piercing process, the dimension design of punch is the main concern, as the hole produced is needed and the workpiece trimmed out from the sheet metal is considered to be waste. In banking process, the dimension of the hole in the die insert is the main thing to be considered, as the final dimensions of the blanked part determines the dimensions of the profile in the die insert. Although both piercing and banking are realized by shearing, the purposes of operations are different.

The second operation is extrusion. In the figure, it shows that the hole made in previous operation is used as the positioning reference shown in the left figure, and the right figure shows the extrusion and its completion status. The last operation in the figure is blanking, which trims down the deformed part from the sheet metal. It can be seen that the earlier described disadvantages in the traditional bulk forming of meso- and microscaled parts can be avoided.

For the process shown in the right figure, it is used to process solid bulk part with different size scales. For the shown preforming process in the figure, it employs shearing process to pierce a hole or profile for positioning or produce a blanked part via blanking. For the second process, the prior-made hole is used as the positioning feature for holding the guider to ensure the alignment between punch and the designed die geometries and further the accurate feeding of sheet metal horizontally. In this step, punch conducts the extrusion deformation of sheet metal in the confined deformation area between punch and die feature. In the third process, the extruded material is blanked, and the deformed bulk solid part is produced.

To achieve the earlier described rationale of progressive and compound forming, a progressive forming system was developed for fabrication of hollow parts, and the schematic operations are shown in Figure 7.4. From the figure, it can be found that the progressive process consists of four stations, including shearing, extrusion, second extrusion, and blanking. In the first step, a hole to be used for positioning is pierced and a cylinder is blanked out. The diameters of punch and die in the shearing process are 0.740 and 0.785 mm, respectively, and the one-side punch-die clearance is 22.5 μm. In the second and third operations, they are all extrusion deformed, and the prior-made hole is used as positioning as the positioning art of punch go into the hole first, and the punch presses the material of the sheet to deform it. In the fourth operation, the extruded geometries are blanked and hollow-based flanged parts are produced.

From the tooling point of view, the tooling material is high-speed steel with the Rockwell hardness (HRC) of 60–62. The experiment was conducted on a materials testing system (MTS) using a load cell, with a maximum capacity of 30 kN. The part was fabricated using a long copper strip with a width of 10 mm and a length of 80 mm. Three lubricated conditions, viz., mechanical oil, lithium grease, and dry friction, were used at the tooling–workpiece interface, such that the influence of lubricant on deformation behavior can be

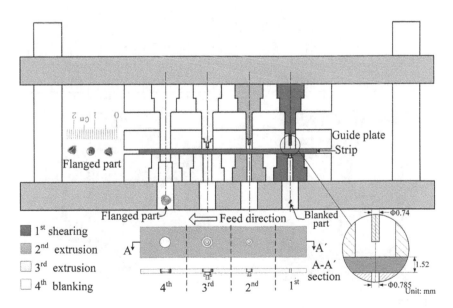

FIGURE 7.4
Schematic illustration of the microshearing process designed for progressive forming [16].

analyzed. The shear speed varying from 0.01 to 0.3 mm/s was used, and the effect of shear speed on the part quality can be explored.

Figure 7.5 presents the schematic illustration of the developed progressive microforming system for production of the flanged part and the forming system can handle shearing, blanking, and extrusion operations. The copper strip with a dimension of $80 \times 10 \times 1.52$ mm was used. It can be seen that there are four processing operations in each stroke. In the first step, the copper strip undergoes shear deformation. A cylindrical part is blanked out and a hole in the sheet metal is pierced, which is used for positioning the subsequent operations. In the second and third operations, the workpiece is extruded in the thickness direction to form multilevel flanged features, and a portion of the material is further pressed into the die orifice. It is noted that the formed part at the first three stages is attached to the sheet metal, which makes it easier for handling and transportation to the next step for blanking. In the last operation, the flanged part is trimmed away from the strip by blanking operation. Considering the machine capacity and mechanical characteristics of each forming operation, a different working procedure is arranged in one stroke via adjusting the punch length of each operation. The punch for blanking the extruded flanged part first contacts the sheet metal. After that, the piercing punch shears the cylindrical part and pierces a hole on the strip. Subsequently, two extrusion processes with the same punch length are conducted simultaneously to form the multilevel flanged characteristic.

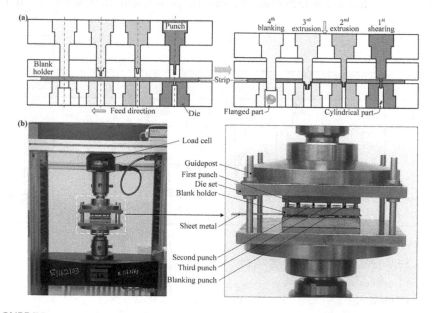

FIGURE 7.5
Developed progressive microforming system: (a) schematic illustration and (b) testing machine and tooling installation [17].

FIGURE 7.6
Progressively microformed cylindrical parts and two double-flanged components [17].

The developed progressive microforming system is shown in Figure 7.5b, and the figure shows its installation in a programmable MTS testing machine with a load cell with the maximum capacity of 30 kN. The total punch stroke is 4.5 mm and the forming speed is 0.01 mm/s. Such a slow forming speed is used to avoid the dynamic effect in forming process. In addition, the local cell, blank holder, guidepost, sheet metal, and the punches at different stations are shown in the figure.

The formed cylindrical and flanged parts are shown in Figure 7.6. It can be seen that the parts can be effectively fabricated using the progressive microforming consisting of shearing, blanking, and extrusion processes directly using sheet metal. The fabricated parts show the efficiency of the process, which can actually overcome the issues such as handling, transporting, and positioning. The dimensional accuracy and the shearing fracture surface quality, however, are other issues to be addressed in this promising meso- and microforming process.

7.3 Shearing Deformation in the Process

The deformation behaviors and dimensional accuracy in meso- and microscaled progressive and compound bulk forming of sheet metals are very critical as they affect the efficiency and performance of the process and further the quality of the deformed bulk components. In this section, the detailed shearing operation and its related deformation behaviors and dimensional accuracy are presented and analyzed in the following.

7.3.1 Shearing Deformation

To explore the deformation behavior in shearing process of the meso- and microscaled progressive and compound bulk of sheet metals, pure copper sheet is selected as the forming material for its excellent formability and conductivity. The copper sheet with a thickness of 1.52 mm was used.

To obtain different microstructures and grain sizes, the materials were annealed with different temperatures and holding times, viz., 500°C for 2 h, 600°C for 2 h, and 750°C for 3 h in an argon gas-filled chamber. In addition, a series of microhardness tests (indentation load: 500 gf) are performed on a microhardness tester (type: FM-700). The material conditions and hardness are shown in Table 7.1, and the microstructures corresponding to different annealed temperatures and flow stress curves are presented in Figure 7.7. In Table 7.1, the hardness was obtained via averaging the six measurement results on per sheet metal. It is found that the material softens with annealing temperature and the deviation of hardness increases with temperature.

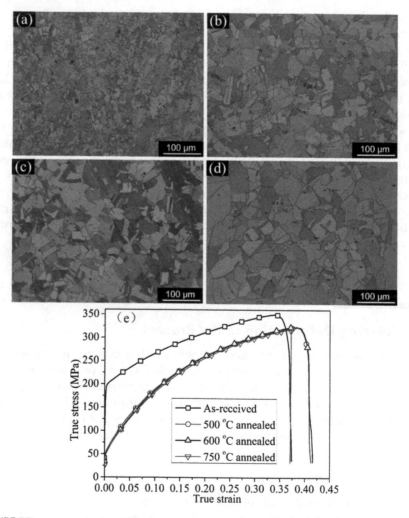

FIGURE 7.7
Microstructures of the testing material: (a) as-received; (b) 500°C, 2 h; (c) 600°C, 2 h; (d) 750°C, 3 h; and (e) flow stress curves in different conditions [16].

TABLE 7.1

Microhardness and Grain Size of the Test Material

Temperature (°C)	As-received	500	600	750
Time (h)	0	2	2	3
Hardness (HV)	80.7±0.5	42.5±0.6	42.4±1.1	42.4±2.0
Average grain size (μm)	11.1	21.9	23.3	37.8

This is due to the random orientation and distribution of grains and the large grain size in the material after high temperature annealing, which leads to an inhomogeneous material behavior.

7.3.2 Deformation Load

To reveal the more detailed material deformation behaviors, deformation load was first investigated by physical measurement and finite element (FE) simulation. For simulation, DEFORM™-3D was used to simulate the micros-hearing process. There are four objects to be considered in simulation, which include punch, blank holder, die, and sheet metal. Except for the sheet metal, all other objects are considered rigid body, while the sheet metal was defined as a plastic body. Different mesh sizes were used for the sheet metal. A very dense mesh of 30 μm was employed for simulation of the fraction of the strip near the deformation zones, and other parts were meshed with a size of 200 μm to reduce computation time. In addition, the normalized Cockcroft–Latham (C&L) criterion was employed to model the fracture behavior, which gives a satisfactory prediction on ductile fracture in shearing process. Figure 7.8 shows the load–stroke curves in microshearing process obtained via experiments and FE simulation. A systematic deviation between the slopes of elastic part of the two curves can be found, which is presumably related to the stiffness of the forming tool and press. Since the load of microshearing is insignificant compared with the capacity of press, the deviation is mainly caused by the compliance of tooling, which should be taken into account by correcting the experimental data. According to the method used in the literatures [18,19], the corrected punch stroke is given as follows:

$$\Delta x_{corrected} = \Delta x_{measured} - \Delta l \quad (7.1)$$

where

$$\Delta l = \frac{F}{K_{tool}} \quad (7.2)$$

where K_{tool} equals to 7,000 N/mm, determined via minimizing the least-square error between the experimental data and simulation in the elastic range of $F=0–300$ N. The corrected load–stroke curve is also shown in Figure 7.8 using this method.

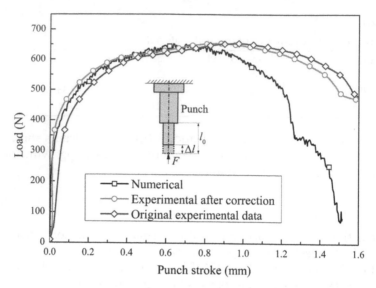

FIGURE 7.8
Correction of the load–stroke curve [16].

To explore the interactive effect of grain size and the lubricated condition on the microshearing process, the load–stroke curves under different grain sizes and lubricated conditions with a constant shear speed of 0.01 mm/s are presented in Figure 7.9. The scatter of shearing load increases with grain size. This is attributed to the variation of material deformation behavior with a small number of grains produced by different annealing temperatures. The grain orientation, boundary, and properties play a big role in deformation when there are only a small number of grains, such as a few grains, in the deformation zone, which leads to a large scatter of the shearing load. From a statistic perspective, fine-grained material contains a large number of grains and the deformation behaves homogeneously as the role the individual grain plays in deformation is not so significant, resulting in a stable statistics result, a homogenous deformation in different directions, and good deformation repeatability.

To further reveal the variation of shearing load versus grain size, the nominal ultimate shear strength σ_s is introduced, which is defined in the following:

$$\sigma_s = \frac{F_{\max}}{\pi d_p t} \qquad (7.3)$$

where $F_{\max}$, d_p, and t are the maximum shearing force, punch diameter, and sheet metal thickness, respectively. In addition, the normalized ratio of die clearance to mean grain size is introduced to represent the interactive effect of the microshearing process designated in the following equation:

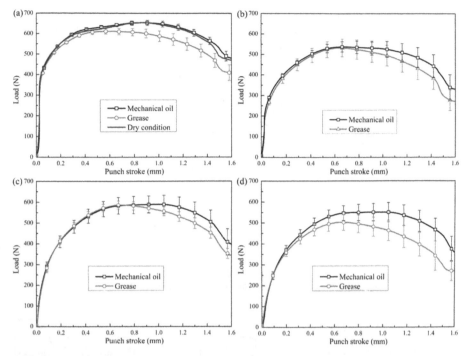

FIGURE 7.9
Load–stroke curves under different grain sizes: (a) 11.1; (b) 21.9; (c) 23.3; and (d) 37.8 (Unit: μm) [16].

$$Z = \frac{c}{d_m} = \frac{d_d - d_p}{2d_m} \qquad (7.4)$$

where c is the punch-die clearance, d_d and d_m are the inner diameter of die and the average grain size, respectively. Figure 7.10 presents the ultimate shear strength versus the ratio of die clearance to grain size. The variation of ultimate shear stress (USS) is mainly related to the interaction of die clearance and grain size. When $c > d_m$, the punch-die clearance has the width magnitude of several grains, and the shear band involves a large number of grain boundaries, as illustrated in Figure 7.11a. When $c < d_m$, there are still several grains in the material thickness direction, as shown in Figure 7.11c. In this case, the deformation of grain matrix takes a leading position in the shearing band, and the grain boundary sliding is still dominant in the thickness direction. When the grain size reaches the same magnitude as the punch-die clearance, the clearance can only accommodate one grain, but there are several grains in the material thickness direction, as shown in Figure 7.11b. The grain boundary sliding and coordination in the clearance width direction reduces to a minimum level. Meanwhile, the grain matrix deformation reaches the lowest magnitude compared with the case of $c < d_m$. The two factors lead to the lowest shearing stress when $c = d_m$.

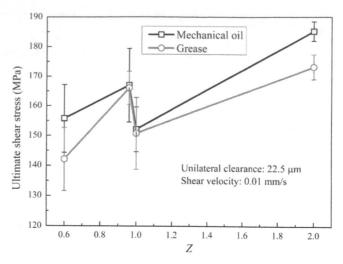

FIGURE 7.10
Effect of the ratio of die clearance to mean grain size on the USS [16].

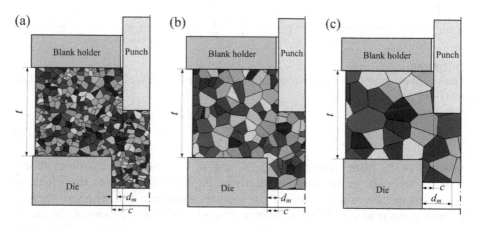

FIGURE 7.11
Schematic illustration of the interactive effect of die clearance and grain size: (a) $c > d_m$; (b) $c \approx d_m$; and (c) $c < d_m$ [16].

In addition, the deformation load shown in Figures 7.9 and 7.10 reveals that the lubricated condition has a significant influence on the deformation behavior in shearing process. In the experiment, mechanical oil and lithium grease were used, and the kinematic viscosities of the two lubricants are 32 and 200 mm²/s, respectively. It is found that the shearing force is reduced by using lithium grease, while the deformation load shows no obvious variation using mechanical oil compared with dry friction condition. According to the lubricant pocket model, the lubricant is entrapped in the asperities between the tool and workpiece. In the so-called closed lubricant pockets, the lubricant is entrapped, making a hydrostatic pressure to bear part of the shearing force. Figure 7.12 presents the comparative analysis between mechanical oil and grease. In the figure, the mechanical oil case is shown on the left side, while the right side represents the grease lubricant. Since the die is vertically positioned, the mechanical oil cannot stay on the inner surface of the die in forming process. As a consequence, the closed pockets fail to develop and will not bear any shearing load, resulting in a similar deformation load to the dry friction condition, as shown in Figure 7.9a. In the case of grease lubricant, the lubricant maintains on the surfaces of the tool and workpiece due to its high viscosity. There are several pockets sealed fully by the surrounding uninterrupted direct contact of the die and cylinder surface, and a hydrostatic pressure p_g develops, which results in the lower deformation load. Nevertheless, the closed pocket shifts constantly during shearing deformation, the instable quantity of closed pockets aggravates the fluctuation of load–stroke curves. Therefore, the lubricant with a low viscosity has a limited role in reducing the deformation load and improving surface quality, while the high-viscosity lubricant is proven to be more effective.

7.3.3 Working Hardening Induced by Shearing

Work hardening is the strengthening of materials by plastic deformation, which resulted from the generation and movement of dislocations within the crystal structure of materials. To evaluate the hardening generation at the rim of the pierced hole, a series of microhardness tests were performed along

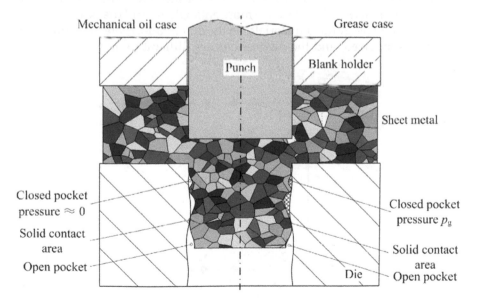

FIGURE 7.12
Schematic illustration of the lubricated effect of mechanical oil and grease [16].

the regions of concern. Due to the different deformation behaviors between the rollover and fracture zone, the microhardness near the two regions was measured, as shown in Figure 7.13.

In Figure 7.14, the variation of hardness depending on the distance from the hole edge with a shear speed of 0.01 mm/s is presented. It is observed that the material hardness is affected by shearing over a distance, which depends on the grain size and the lubricated condition. When the fine-grained material is used, the hardness is generally high compared with the materials with coarse grains. Figure 7.14 show the high hardness of the material with the average grain size of 11.1 μm compared with that of the materials with coarse grains such as the average grain size of 37.8 μm. This follows the great hardening effect of the fine-grained materials.

In additon, for the hardness increase induced by shearing, there is a subtle increase for the fine-grained material caused by a small amount of deformation at the rim of the pierced hole. For the coarse-grained materials, however, the hardness increase is about 44% compared with that of the original material. For the former, it is due to a large number of fine grains involved in the deformation, and the deformation can be facilitated by different fine grains in the shearing process, while for the latter, however, the grains are large and the shearing needs to break down the large grains, and the shearing deformation goes through grain boundaries. It is well known that the grain boundaries act as an obstacle, disrupting the dislocation glide along the slip planes and thus the dislocations cannot easily cross the grain boundaries due to the change in the direction of slip plane and the atomic disorder at grain boundaries. For the materials with large grains, it is not easy for the deformation to go across the grain boundaries, which leads to more entanglement of dislocations in the pileup. For the fine-grained materials, on the other hand, there are more grain boundaries involved, and the dislocations are difficult to move through the

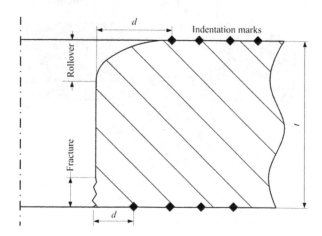

FIGURE 7.13
Distribution of indentation marks in the vicinity of pierced hole [16].

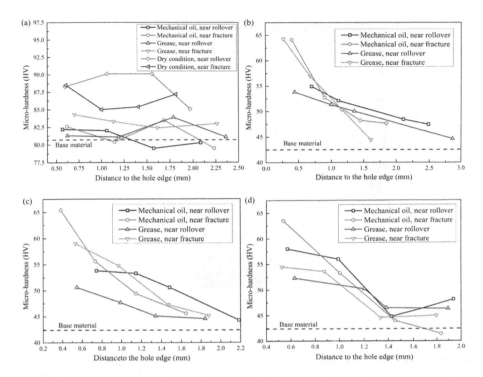

FIGURE 7.14

Microhardness profile near the pierced surface under different grain sizes: (a) 11.1, (b) 21.9, (c) 23.3, and (d) 37.8 (Unit: μm) [16].

grain boundaries, resulting in an increment of dislocation density. Therefore, the shearing process with coarse-grained material has a greater work hardening compared with the fine-grained material. Meanwhile, it can be seen from Figure 7.14a that the hardness under dry friction condition presents a significant high value compared with the lubricated condition. This is because the material is subjected to severe deformation again, with the punch extracting from the strip, which further facilitates work hardening. Compared with the regions close to rollover zone, the work hardening is higher at the surface near die entrance due to the more shearing deformation lasting until the occurrence of fracture. It is also noticed that the hardening is lower with grease lubricant for a fixed distance to the hole edge, revealing that the lubricant with high viscosity reduces hardening accumulation.

7.3.4 Dimensional Accuracy

The forming quality of the blanked part via shearing is mainly affected by material properties, punch-die clearance, material thickness, and shear speed. The cylindrical part is characterized by the distinct zones, as shown in Figure 7.15. In the figure, h, d_b, h_r, h_s, h_f, and h_b are billet height, billet

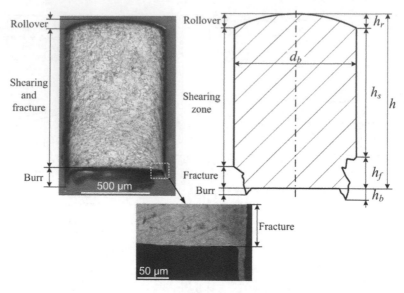

FIGURE 7.15
Different zones of the sheared micropart [16].

diameter, rollover length, shearing zone length, fracture length, and burr height, respectively. To identify the forming quality of the profile of both the blanked part and the pierced hole after shearing, five specimens in each condition were prepared for micrographic analysis and measurement. The surface morphology of the cylinder and punched hole was observed with scanning electron microscope, while the profile of the formed part was measured using microscopy technique.

The dimensional accuracy and surface finish of the blanked part and pierced hole were examined to characterize the forming quality in the customized shearing process. Figure 7.16 shows the change of cylinder height h and diameter d_b with average grain size using different lubricants. It is found that the length of cylindrical part decreases with grain size, while the part diameter reaches the maximum and then decreases. Chan and Fu [20] found that the length of the cylinder was decreased with grain size, which is attributed to a large amount of lateral material flow taking place in the coarse-grained material. Furthermore, the part diameter reaches the maximum when the punch-die clearance is close to grain size. The variation of part diameter is associated with the interactive effect of die clearance and grain size shown in Figure 7.11. For the fine-grained material, there are more grains involved in the shearing process, and thus there are more grains to fill the die cavity. When the grain size is greater than die clearance, the material is more difficult to be pushed into the die cavity, and the shape of cylindrical part becomes irregular. When the grain size equals the punch-die clearance, more grains slide along the die edge, and the entered grains can exactly occupy the whole die

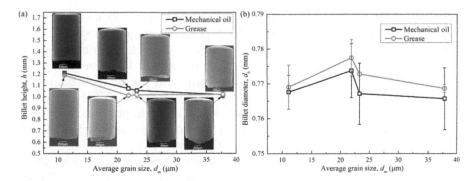

FIGURE 7.16
Dimensions of the blanked part: (a) height and (b) diameter [16].

cavity, resulting in the peak value of billet diameter. In addition, the billet diameter with grease lubricant is larger, which is because the material tends to be attached to the side wall of die opening with low frictional force.

To further investigate the variation of the dimension of the cylinderical part, the volume of the sheared part was calculated according to the observed appearance, as shown in Figure 7.17. It can be seen that the part volume is reduced with grain size due to a samll amount of deformation in shear band with coarse-grained material. The frictional resistance of grease is much smaller than that of mechanical oil, bringing about a large amount of deformation involved in the lateral direction. Therefore, both the sheard part length and volume decrease by using grease lubricant.

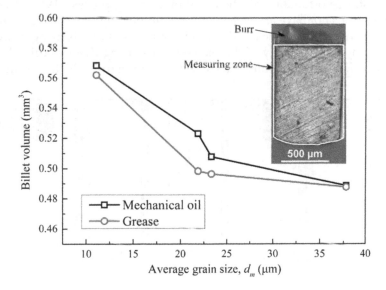

FIGURE 7.17
Variation of the billet volume versus grain size [16].

The surface of the blanked cylinder can be divided into four zones, including rollover, shearing, fracture, and burr. The depth percentage of these regions is illustrated in Figure 7.18. It can be found that both the rollover and burr grow with grain size, while the shearing and fracture zones decrease with grain size. The rollover is caused by the initial elastic and plastic deformation as the punch touches the material and travels a small distance. The burr occurs when the cracks starting from the punch to die edge do not match, and the material will be dragged and torn through the die edge, resulting in a ragged edge at the end of the cylinder. The length of burr is closely related to the interaction of grain size and punch-die clearance, and reaches a peak when the grain size equals to die clearance. The fracture is associated with the growth and coalescence of microvoids. Mori et al. [21] revealed that the fracture regions became considerably large when a hole of 10 mm was punched on the ultrahigh strength steel sheet with the thickness of 1.22 mm. When the dimension of the pierced hole is reduced to submillimeter, the number of grains and the activated slip systems involved in the deformation zone decrease. In such a way, the fracture phenomenon is not obvious in micro-shearing process, whereas the shearing becomes a dominated deformation pattern. The rollover and burr on the pierced part would significantly affect the positioning accuracy and forming quality in the subsequent operations, while the fracture could influence the geometric accuracy of the final part. Furthermore, the lubricated condition has a significant effect on the distribution of the earlier zones. Compared with the mechanical oil, the shearing and burr increase, and the fracture zone decreases by using grease lubricant. The friction is relatively large when using mechanical oil, which leads to a large plastic deformation and the delay of shearing process. While the rollover zone does not have an obvious change, indicating that the lubricant can only have an impact on the middle and later stages of the shearing process.

The surface morphology of the pierced holes using the as-received material under different lubricated states is shown in Figure 7.19. It is found that

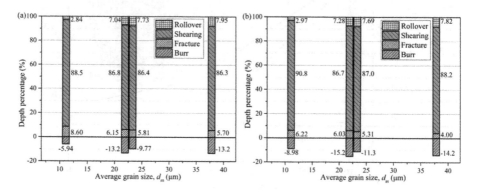

FIGURE 7.18
Dimensions of different zones of the blanked part under different lubricated conditions: (a) mechanical oil and (b) grease [16].

FIGURE 7.19
The pierced holes under different lubricated conditions: (a) mechanical oil; (b) grease, and (c) dry friction [16].

the surface near the rollover zone becomes rough, and some rags can be seen at the rim of the pierced hole under dry friction condition. It could be attributed to the secondary damage as the punch extracted from the part under high friction force. The surface around the hole is more smooth with the grease lubricant, and the punched hole is more regular. In fact, the grease that acts as a protective film during the forming process, leading to a more burnished surface.

In addition, the grain size and lubricated condition also affect the microstructures and surface finish of the blanked part. The microstructures of the blanked part under different conditions are shown in Figure 7.20. It can be seen that the shape of the cylinder part becomes irregular with increase of grain size. That is because the dominant individual grain strengthens the inhomogeneous deformation and the discordance between the grain slip direction and the shearing direction.

The evolution of microscopic structure of the blanked part using the material annealed at 600°C is shown in Figure 7.21. In the figure, the stress states during the shearing process are also presented via FE simulation. According to the variation of grain size and stress state, the microstructure of the cylinder can be divided into four zones such as zones I, II, III, and IV. Among these zones, the less-deformed zone, zone I, is defined as the dead zone by Ghassemali et al. [22]. The highly elongated and rotated grains close to the shear bands can be seen in zone IV. When the elongated grains extend to the center section, an arched band, zone II, is compressed to be formed. The average grain sizes of zones I and II in the punching direction are 14 and 4.4 µm, while the values become 15 and 18.9 µm perpendicular to the shearing direction, respectively. This indicates that the material in zone II is subjected to the tensile force in the horizontal direction, which is induced by the restriction near the edge of moving punch. The tensile stress in zone II transmits the punching force to the punch edge to separate the material from the sheet metal. The shape of zone II is associated with the different flow velocities between the center and both sides of the blanked part. Furthermore, the bandwidth of zone II is related to the original grain size. As the fine-grained material is used, the grains are distributed evenly in zones II and III, and

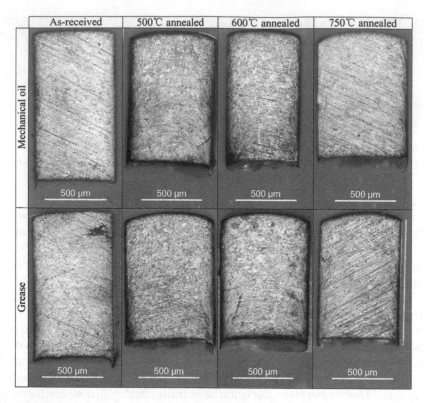

FIGURE 7.20
Microstructure of the blanked part under different states [16].

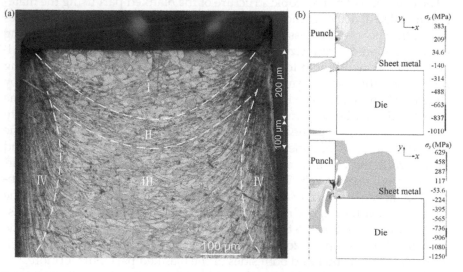

FIGURE 7.21
Evolution of microstrcture of the blanked part [16].

thus the compressive band is not obvious. The stress state of zone I is under triaxial compressive stress, which is attributed to the fact that zone II hinders the material in zone I to flow on both sides. The grain size of zone III changes from 8.7 at the top to 18.2 μm near the rollover gradually. The stress state of zone III is the same as that in zone II, but the tensile stress is lower, resulting in a gradient distribution of grain size.

In conclusion, the quality of the blanked part and the pierced hole in terms of dimensional precision and surface integrity is mainly determined by grain size, die clearance, and lubricated condition. From mass production perspective, both grain size and material thickness should be considered in the design of punch-die clearance. Meanwhile, the fine-grained material and high viscosity lubricant are preferred to improve forming quality. Besides, the shear speed is also a significant factor associated with the forming quality and production efficiency, which should be determined based on the comprehensive analysis of quality and productivity.

7.3.5 Speed Affected Deformation Behavior and Part Quality

In mass production, the forming speed should be considered. An ideal shear speed is needed to balance the surface quality, tool wear, and manufacturing efficiency. To investigate the effect of shear speed on forming performance, four velocities including 0.01, 0.05, 0.1, and 0.3 mm/s were used to shear the material annealed at 750°C under grease lubricant. Figure 7.22 shows the load–stroke behavior under different velocities. In the figure, the deformation load grows with shear speed, while the scatter of the load curve

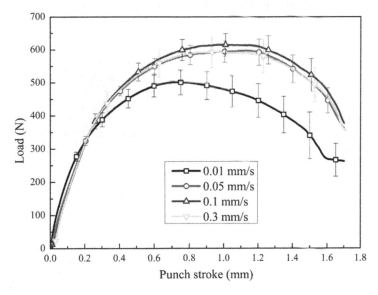

FIGURE 7.22
Load–stroke curves under different shear speeds [16].

decreases. This is because the work hardening enhances with deformation rate. When the shear speed is larger than 0.05 mm/s, the variation of deformation load is not obvious.

To study the effect of shear speed on work hardening, the microhardness in the areas near the pierced hole was measured. Figure 7.23 shows that the hardness around the hold edge is larger than that of the remaining region due to the hardening accumulation effect of shearing deformation. When the velocity is greater than 0.05 mm/s, there is no significant influence on work hardening and deformation load.

Figure 7.24 shows the surface finish of the fabricated billets using different shear velocities. The surface of the cylindrical part becomes rough at high shear speed. Furthermore, the geometries of the different zones change with the deformation velocity, and the measured dimensions of the blanked part are presented in Figure 7.25. It is found that the proportions of burr and rollover zone grow, and the profile becomes irregular at both sides with increase of shear speed. This is due to the fact that the different deformation rates of material around the punch bottom and die edge, and the generated turbulent material flow under high deformation velocity may result in an irregular configuration. With high punching velocity, the velocity gradient of the material flow enhances along the shearing direction, and the working material sticks to the die edge. The delayed deformation material around the die edge makes a longer burr zone. In addition, the high shear speed can generate not only a greater deformation force but also a larger friction. The enhanced friction force at high velocity will favor the compressive deformation of the billet, resulting in a shorter length and a larger diameter of the blanked part, as shown in Figure 7.21. The high shear speed also causes a

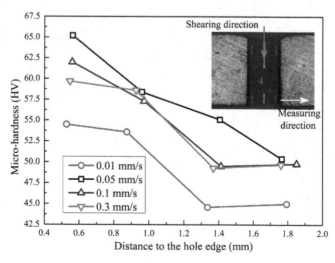

FIGURE 7.23
Microhardness profile around the pierced hole [16].

FIGURE 7.24
Surface characteristic of the blanked part using different shear speeds [16].

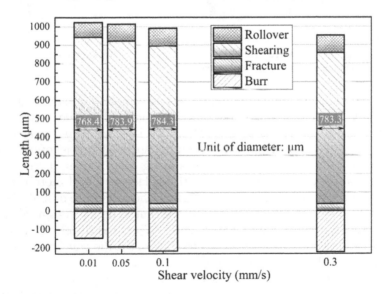

FIGURE 7.25
Dimensions of the blanked part using various shear speeds [16].

large plastic deformation and a reduction of shearing zone, while the change of fracture zone is not obvious.

It can be found that several surface breakage defects appear under the shear speed of 0.3 mm/s, as shown in Figure 7.26a. The primary cause of

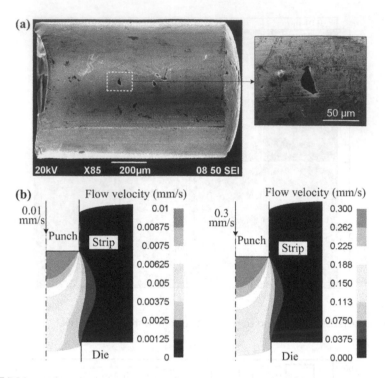

FIGURE 7.26
Surface breakage on the sheared part and its causes: (a) surface breakage and (b) material flow characteristics under low and high shear speeds [16].

the surface cracking is the inhomogeneous deformation rate throughout the shear band. The higher shear speed and friction will create a more favorable condition for surface cracking. The material flow in the shearing process obtained via FE simulation is shown in Figure 7.26b. It is found that the deformation velocity gradient decreases along the shearing direction. The center region experiences more material displacement and more turbulent flow than the areas toward the outer region. The difference of material flow between the region around die edge and the central region is significant, and this difference enhances with shear speed. When the difference reaches a certain level, the outer material tends to stick to the die edge, and the material breakage will occur in terms of surface crack.

The shear speed greatly affects not only the forming quality of the blanked part but also the pierced holes. Figure 7.27 shows the surface morphology and profile characteristics of the pierced holes under different shear velocities. The flash appears at the hole edge under a shear speed of 0.3 mm/s, which is attributed to the turbulent flow of surface grains as the punch pushes the material into the die orifice at high velocity. From the profile characteristic, the fractions of rollover and burr increase with shear speed, which would seriously affect the positioning accuracy and product quality in subsequent

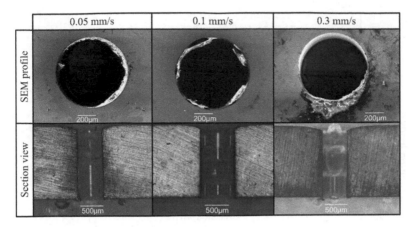

FIGURE 7.27
Surface characteristic of the pierced hole under different shear speeds [16].

operations. Therefore, the shear speed should be controlled within a limited range to prevent defect formation. In this study, the shear speed should be limited to below 0.1 mm/s to balance the surface quality, tool wear, and productivity.

7.4 Progressive and Compound Deformation in the Process

For the progressive and compound bulk forming process directly using sheet metal shown in Figure 7.5, the deformation and fracture behaviors and the product quality of the deformed part are articulated in the following section.

7.4.1 Deformation Behavior

The deformation behavior discussed in this section covers two parts, viz., deformation load and material flow behaviors. They are discussed in the following:

7.4.1.1 Deformation Load

In this progressive forming process, the deformation load comes from different processing operations, including shearing, extrusion, and blanking. Figure 7.28 shows the load–stroke curves in the progressive forming process under different conditions. Considering the operation sequence as mentioned earlier, three forming stages in the progressive forming process are identified. In the first stage, the flanged part is subjected to blanking deformation, and the final part can be obtained after the punch stroke reaches 0.5 mm. In the second stage, the punch stroke from 0.5 to 2.0 mm stands for the shearing

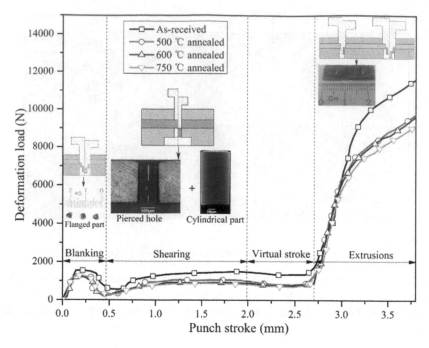

FIGURE 7.28
Load–stroke behaviors under different forming states [17].

process, which is used to fabricate the cylindrical part and the pierced hole. After the fracture occurs in shearing process, the remaining force is caused by overcoming the friction, as the blanked part slides along die outlet. In the last stage, viz., the punch stroke ranges from 2.7 to 3.8 mm, and the curve represents the two-step extrusion process to form the multilevel flanged features that are conducted simultaneously. It is noted that the virtual stroke between shearing and extrusion processes, viz., punch stroke varying from 2.0 to 2.7 mm is designed to adopt different strip thicknesses. It can be found that the deformation load of the as-received sheet metal is larger than that of the annealed ones. This is attributed to the fact that the as-received material was hardened in the rolling process, in addition to grain boundary strengthening effect. However, the number of grains reduces over the material thickness with annealing temperature, resulting in inhomogeneous deformation and large localized deformation with coarse-grained material.

7.4.1.2 Material Flow Behavior

From the perspective of material flow behavior, the flow pattern of the progressive form is first considered, which is closely associated with the micropart quality in terms of surface integrity and defect formation. Thus, the understanding of the material flow behavior is crucial to optimize the process.

Figure 7.29 shows the simulated material flow of the second forming operation. It can be found that a considerable part of the material flows laterally at the beginning of the punch penetration, resulting in a short extrudate. As the punch stroke grows, more and more materials flow towards the die orifice, leading to a rapid increase in extrudate height. This indicates that there is a critical zone to divide the material flow in two directions. The inner material flows toward the die cavity to contribute the length of extrudate, while the outer material flows outwards. The balance of the two flows depends on the resistance in the corresponding flow direction. In addition, the length of the extrudate with fine-grained material in the second operation is about 300 μm longer than that with coarse-grained material when the punch stroke is 1.0 mm, which is induced by the different flow rates in the extrusion direction. For a given part with a specified length of the extrudate, the forming system should be optimized to increase the amount of forward material flow and reduce the material waste. From this perspective, the fine-grained material is preferred in the progressive microforming.

Figure 7.30 shows the microstructures of the extruded regions in the two continuous extrusion operations. It can be observed that the shear band extends from the punch edge to the die orifice. The grains at the shear bands underwent a severe plastic deformation, and the effective strain is thus highly localized in this region. For both second and third operations, the

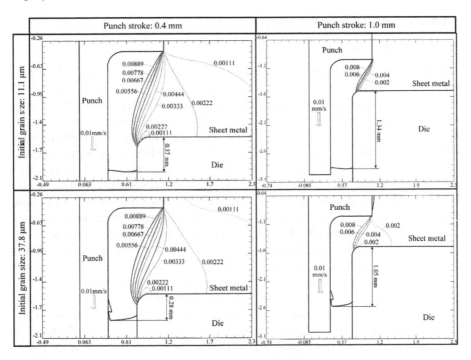

FIGURE 7.29
Distribution of flow velocity in the second operation calculated by FE simulation [17].

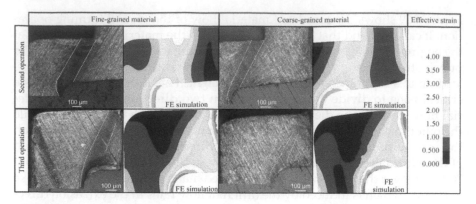

FIGURE 7.30
Shear bands in the extrusion processes [17].

shear band in the fine-grained material is larger compared with the coarse-grained material, which is related with the different flow characteristics. The shear band involves a large amount of grains to participate the plastic deformation using fine-grained material, resulting in more severe deformation in the shear band.

7.4.2 Ductile Fracture Behavior

7.4.2.1 Fracture Surface

The ductile fracture of material is caused by the progressive degradation of material stiffness when the plastic deformation reaches a certain limit. In the progressive deformation scenario, the ductile fracture is easier to come into being due to the repeated deformation cycle and severe damage accumulation. In this research, the ductile fracture phenomena were mainly induced by individual operation and interactive effect of the progressive strain cycle. Figures 7.31 and 7.32 show the fracture surfaces of pure copper sheet observed by the scanning electron microscope after the shearing and blanking operations, respectively. It is found that there were some dimples on the blanked surfaces of the flanged part, which were caused by void growth and nucleation. However, no obvious fracture surface was observed on the punched billet with a diameter of 0.74 mm, especially for the coarse-grained material. That indicates that the fracture phenomenon is more likely to occur as the geometrical dimension increases to meso- or macroscale in the shearing process. It is because the number of grains and grain boundary involved in the deformation region reduces with the decrease of workpiece dimension. In such a way, the microvoids tend to occur not only in the microscaled part due to the voids but also at the grain boundaries or inclusions. The fracture mechanism based on the initiation, growth, and coalescence of microvoids results in a rough surface finish in

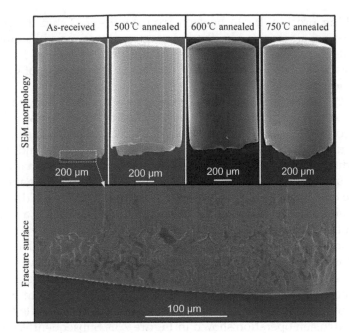

FIGURE 7.31
The fracture surface on the cylindrical part during shearing operation [17].

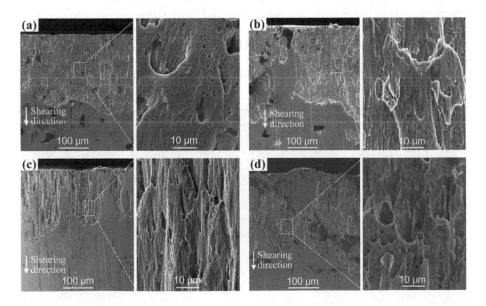

FIGURE 7.32
The fracture surfaces on the flanged part during blanking operation: (a) as-received; (b) 500°C; (c) 600°C; and (d) 750°C annealed [17].

the blanking process. It can be seen from Figure 7.32 that the parabolic-shaped microvoids occur on the fracture surfaces, and the number of microvoids decreases with grain size. In addition, the size of fracture surface shows an increasing tread with the decreasing number of grains over the thickness of the workpiece.

Figure 7.33 shows cross-section photographs across the flanged part. It can be observed that the extruded surface in the second operation was fractured with the corrugated surfaces caused by void growth. The fracture induced by strain accumulation weakens with the increase of grain size, and there is no obvious ductile fracture for the material annealed at 750°C. That is because there is a small amount of grains involved in extrusion deformation; meanwhile, the deformation degree and strain accumulation effects decrease as a coarse-grained material is used. Another fracture model was a microcrack located at the sidewall of the flange, as shown in Figure 7.34. The microcrack results from the accumulative effects of second extrusion and blanking operations. Moreover, the microcrack phenomenon was only observed on the as-recieved material. The material was subjected to severe strain accumulation in the second extrusion stage, and the strain-induced fracture was generated on the extrudated surface, which is the same as the ductile fracture in the first extrusion operation shown in Figure 7.33. Subsequently, the second flanged feature was blanked out from the sheet metal, and the fractured surface in the extusion stage underwent extusion and shearing deformations once again. When the fractured surface was cut off from the copper sheet, the microcrack emerged on the flange of the final workpiece. The internal characteristic of the microcrack was observed from the cross-section view, as shown in Figure 7.34. It can be seen that the microcrack extends to appropriate 200 μm from the part edge, indicating that the crack was formed before the blanking stage.

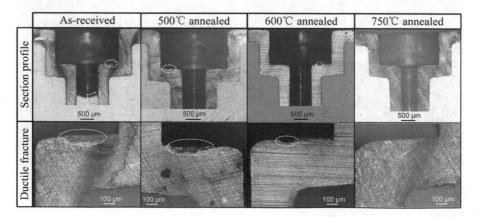

FIGURE 7.33
The fracture surfaces on the flanged part during the second forming operation [17].

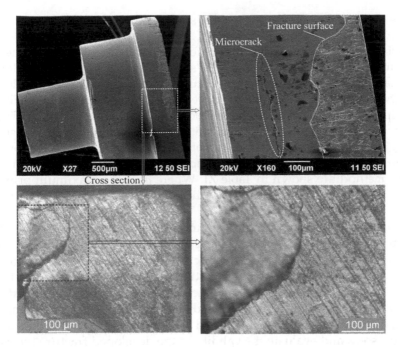

FIGURE 7.34
Microcrack on the flanged part induced by accumulative effect [17].

7.4.2.2 Analysis of Ductile Fracture

In sheet metal forming processes, ductile fracture occurs due to the nucleation, growth, and coalescence of microvoids. Many fracture criteria have been proposed to predict the ductile fracture, which are classified into coupled and uncoupled categories. The coupled fracture criteria incorporated the damage accumulation into the constitutive equations, and thus allow the material yield surface to be modified by the change of the damage-induced density. However, it is difficult to determine the material constants accurately in the coupled criteria. On the contrary, the uncoupled ones have been widely used due to their simpler formulation and easier calibration, despite their limitation in representing the deterioration of the damaged materials. Thus, several uncoupled criteria were chosen and calibrated to investigate their validity and reliability in microforming in this research. In these criteria, damage accumulation is formulated empirically or semiempirically in terms of the macroscopic variables closely related with fracture initiation and propagation, and they are designated in the following:

$$\int_0^{\bar{\varepsilon}_f} f(\sigma, \bar{\varepsilon}) d\bar{\varepsilon} \geq C_c \tag{7.5}$$

where $\bar{\varepsilon}_f$ and $\bar{\varepsilon}$ are the critical plastic strain and equivalent strain, respectively, in the deforming body, and C_c is the critical damage value. The material constant C_c in fracture criterion can be calibrated by the method combining uniaxial tensile test and FE simulation [23]. In this methodology, the maximum principal strain at rupture is first obtained via tensile test, and then the tensile deformation process is simulated until the maximum principal strain of a certain element reaches the experimental value. Finally, the stress–strain relation of the element is substituted into the damage formula, and the material constant is obtained via integration, as shown in Figure 7.35. The material constants under different conditions determined by the method are shown in Table 7.1. Where σ_1 is the maximum principal stress, σ_m is the hydrostatic stress, $\bar{\sigma}$ is the equivalent stress, and $C_1, C_2, C_3, C_4,$ and C_5 are material constants.

In macroscale forming, ductile fracture behavior has been widely studied, and its mechanisms have been well explored. Many ductile fracture criteria have been developed to predict the fracture behavior. In progressive microforming, however, ductile fracture is still a new issue and has not yet been fully investigated due to the unique fracture behavior. This is quite different from those in macroforming or individual forming process due to the size effect, deformation inhomogeneity, and accumulative effect. It is thus necessary to assess and evaluate the validity of the developed fracture criteria in progressive microforming. The applicability of various uncoupled fracture criteria was thus evaluated via progressively forming the flanged part. The integral values of $\sigma_1, \sigma_m, \bar{\varepsilon}_f, \bar{\sigma},$ and $\bar{\varepsilon}$ were calculated for all elements and each time step. The first element at which the integral value reached a material

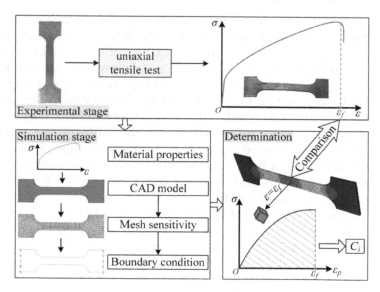

FIGURE 7.35
Determination of the material constants in fracture criteria [17].

TABLE 7.2

Material Constants for Different Fracture Criteria under Different States [17]

Fracture Criteria	Formula	Damage Constants			
		As-received	500°C	600°C	750°C
C&L [24]	$\int_0^{\bar{\varepsilon}_f} \sigma_1 \, d\bar{\varepsilon} = C_1$	104.90	93.37	94.95	84.83
Normalized C&L [25]	$\int_0^{\bar{\varepsilon}_f} \dfrac{\sigma_1}{\bar{\sigma}} \, d\bar{\varepsilon} = C_2$	0.35	0.39	0.39	0.37
Brozzo [26]	$\int_0^{\bar{\varepsilon}_f} \dfrac{2\sigma_1}{3(\sigma_1 - \sigma_m)} \, d\bar{\varepsilon} = C_3$	0.35	0.39	0.39	0.36
Ayada [27]	$\int_0^{\bar{\varepsilon}_f} \dfrac{\sigma_m}{\bar{\sigma}} \, d\bar{\varepsilon} = C_4$	0.12	0.13	0.13	0.12
R&T [28]	$\int_0^{\bar{\varepsilon}_f} e^{\frac{\alpha\sigma_m}{\bar{\sigma}}} \, d\bar{\varepsilon} = C_5$	0.55	0.64	0.62	0.60

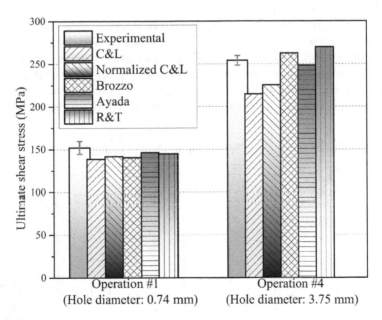

FIGURE 7.36
Prediction of USS in shearing and blanking operations using different criteria [17].

constant was defined as the fracture point. The USS was compared between the simulation results with ductile fracture criteria and those of experiments, as shown in Figure 7.36. It can be seen that the USS predicted by all the selected fracture agrees well with the experimental stress in operation 1, viz.

shearing operation. In operation 4, however, there is a comparative deviation between the simulations and the experiments, except Ayada model. That may be the result from the damage accumulation effect in the last forming stage induced by previous operations. Therefore, the ductile fracture criteria developed in macroforming or individual forming operation are not fully accurate in progressive microforming due to the size effect, work hardening, and damage accumulation.

Figure 7.37 shows the size of the various zones on the punched profile predicted by the selected uncoupled criteria in the shearing operation of pure copper annealed at 500°C. It can be observed that the predicted heights of the rollover zone were in good agreement with the experimental results for all of the fracture models. However, the numerical prediction of the size of the burr area was fairly remote to the experimental results, and the prediction of burr formation was absent for all the fracture criteria. Indeed, the experimental burr accounted for a considerable proportion of the whole profile. In addition, the numerical prediction of the size of fracture zone did not correspond to the experimental ones, except for Ayada criterion. The other criteria, including C&L, normalized C&L, Brozzo, and Rice & Tracy (R&T) models, were unable to predict the fracture initialization and location correctly in the microshearing case.

Figure 7.38 shows the values of the height of fracture zone and the inclination angle predicted by the uncoupled fracture models in the blanking stage of the copper sheet annealed at 500°C. In this figure, h_f is the height associated with the fracture region, and φ is the inclination angle of the flange. It is noted that the C&L model had a good prediction of the value of h_f in the blanking stage, but was far from the experimental data for the prediction of

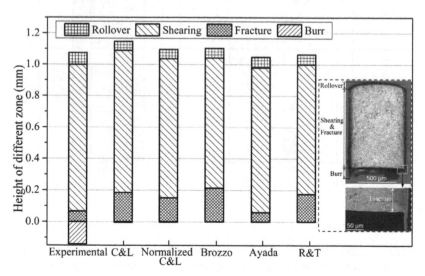

FIGURE 7.37
Evolution of the cylinderical part profile under different fracture criteria [17].

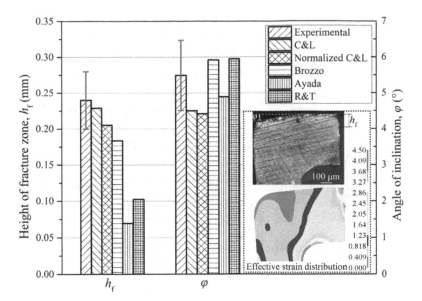

FIGURE 7.38

Evolution of h_f and φ in the blanking stage under different fracture models [17].

fracture zone in the shearing operation, as shown in Figure 7.37. On the contrary, Ayada model was in good agreement with experimental result for the microshearing operation but was not predictive for the macroblanking stage. This can be explained by the fact that the domain of validity of the material constant C_i in the uncoupled models is limited by the range of workpiece size used for its identification. In addition, the inclination angle of the flange reduces the part quality, which is associated with the punch penetration at fracture initiation. It can be seen from Figure 7.38 that the predicted value of φ is less accurate for all fracture criteria. This is because the inclination angle is affected by the comprehensive effects of material state, grain size, die dimension, and fracture behavior. The uncoupled models neglect the effect of damage on the yield surface of the used material, leading to poor prediction precision for the parameters influenced by integrative factors.

For all of the uncoupled criteria selected in this research, the normalized damage values at the punch-cutting edge in microshearing process increased monotonically with a larger equivalent plastic strain, which conforms to the fact that the likelihood of ductile fracture increases with plastic deformation, as shown in Figure 7.39. In addition, the damage value of the C&L model increased faster among all of the criteria compared.

Figure 7.40 shows the variations of damage parameters versus punch penetration for the material annealed at 500°C in the blanking stage. In this figure, U represents the percent of the punch penetration to the thickness of the workpiece flange, H_4. It can be noted that the plastic strain is localized in the shear zone due to the displacement of the punch. The damage increases locally to a

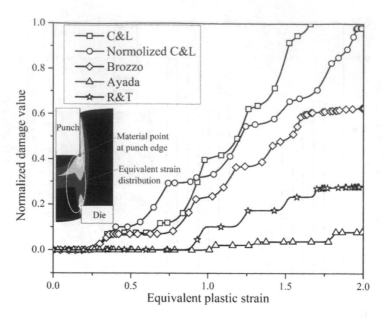

FIGURE 7.39
Damage values of punch-cutting edge against the equivalent plastic strain in microshearing operation [17].

sufficiently large magnitude to induce internal failure of the material and consequently leads to fracture along the thickness direction. The prediction of the location at which crack initiation occurs depends on the used fracture model. Both Ayada and R&T models possess a considerable punch penetration at fracture initialization, while the predicted value performed by C&L is lowest (7.8%). All the selected uncoupled fracture criteria predict that the crack initiates at the punch cutting edge and propagates in the direction of the contact between the sheet metal and die orifice. However, the damage processes are quite different with various fracture models. Using C&L, normalized C&L, and Brozzo criteria, the fracture starts to grow at the punch-cutting edge and propagates towards another crack created at the die orifice. The two cracks propagate towards the center of the sheet metal, resulting in a total fracture of the material.

In addition, the numerically predicted fracture location agrees well with the experimental observations shown in Figure 7.38, with C&L and normalized C&L models. Nevertheless, Ayada and R&T criteria revealed a different fracture pattern, where the crack did not propagate at the die cutting edge, leading to a small fracture zone on the workpiece. The simulative results predicted by Ayada and R&T models were not in agreement with the experimental ones in the blanking stage. In addition, the punch penetration at the complete separation point is closely associated with the speed of fracture propagation. The propagation velocity of the damage predicted by C&L is faster than other fracture criteria, leading to the total fracture at a small punch penetration of 29.2%. However, Ayada and R&T models predict the

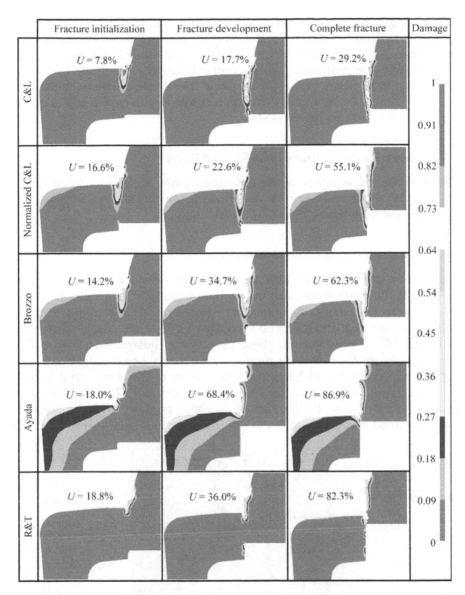

FIGURE 7.40
Damage distribution of different stages of punch penetration in the blanking operation of 500°C annealed material [17].

complete ductile fracture at 86.9% and 82.3% of punch penetration, respectively. The reason for this inability in the case of Ayada and R&T models is the neglect of the largest principal stress in their formulations to capture the localized deformation and damage accumulation. In the C&L, normalized C&L, and Brozzo models, in contrast, the damage location in the blanking operation can be properly predicted, because the key influential factors including hydrostatic stress and the largest principal stress are considered reasonably. Although some of the uncoupled criteria can predict fracture location well in microshearing or macroblanking individually, none of them can predict the fracture behavior accurately for all the operations in the progressive forming because each criterion is limited by the applicable scope of workpiece dimension and stress state.

7.4.3 Dimensional Accuracy of Deformed Parts

The dimensional accuracy of the deformed parts is closely related to the deformation mode and deformation behaviors. In this section, dimensional accuracy and undesirable deformation are analyzed.

7.4.3.1 Undesired Progressive Deformation

Undesired deformation often occurs in the progressive microforming process due to the strain accumulation and interactive effects of each operation. According to the different deformation mechanisms, the irregular geometric defects of the flanged part can be identified in three regions, as illustrated in Figure 7.41. The reasons for these defects are illustrated in Figure 7.42. The features in Zone A were produced by the last two operations, viz., the second extrusion and blanking. The defects in this area result mainly from the blanking operation, as shown in Figure 7.42b. The blanking process itself easily causes the burr and fracture at sidewall, especially when a coarse-grained material is utilized. On the other hand, the stepped feature

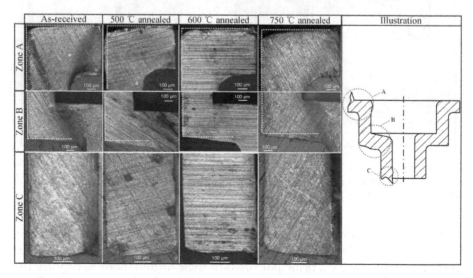

FIGURE 7.41
Geometrical features of the flanged part after progressive deformation [17].

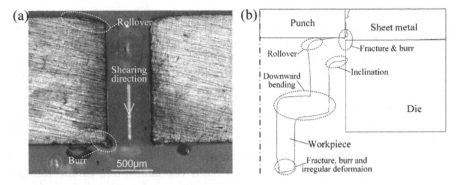

FIGURE 7.42
Reasons for the undesired progressive deformation: (a) effect of shearing operation and (b) effect of blanking operation [17].

extruded by the third operation was warped during the blanking process because there is no container in the blanking process to constrain the formed geometries. The defects in zone B are mainly the inclinations of the stepped features, which are closely associated with the influence of the final blanking operation. When the flanged part is blanked, the deformation not only occurs in the shearing zone but also interferes with the flanged features that were formed by previous stages. Irregular deformation in zone C, located at the bottom of the part, is caused by the first two operation, viz., shearing and extrusion processes. The defects in zone C mainly contain the tortuous geometry and burr, which were induced by the interactive effects of the first two operations. At the beginning of the process, the shearing operation leads to a rollover and burr on the sheet metal, as shown in Figure 7.42a. The burr on the sheet metal underwent extrusion deformation in the next working procedure, which contributes to the rag and irregular deformation in zone C.

It can be also observed from Figure 7.41 that the undesired progressive deformation grows with the increase of grain size. The serious inclination of the flanged features, irregular shape, and the obvious burr can be found for coarse-grained material. That is due to the fact that the size and the orientation of individual grains affect material flow behavior significantly when the characteristic dimension is in the order of grain size. In addition, the material stiffness and strength reduce at a high annealing temperature, which contributes to the further deformation of the formed features. To describe the effects of grain size on the part quality quantitatively, the dimensional accuracy of the feature sizes should be analyzed.

7.4.3.2 Dimensional Accuracy

Dimensional accuracy is, in particular, an important factor in micropart development, since it directly affects the production quality and performance. In progressive microforming, the dimensional accuracy is mainly affected by material property, microstructure, and process variables. Compared with macroscale forming, the meso/micro-scale forming processes introduce new factors that increase the processes complexity such as small dimensions of the parts and inhomogeneous deformation. These are characterized by some common or specific phenomena that affect the dimensional accuracy of the fabricated parts. To evaluate how the annealing condition and scaling factor affect the final part quality, the geometries of the fabricated parts under different conditions were measured. The dimensional variation of the two-level flanged parts is shown in Figure 7.43.

The desired value of D_1 is 3.75 mm, D_2 is 2.2 mm, D_3 is 0.74 mm, D_4 is 1.5 mm, and D_5 is 3.0 mm. It is found that the diameters of the flanged part, viz. D_1, D_2, D_3, D_4, and D_5 are exactly equal to the desired geometries for both fine- and coarse-grained materials, in contrast to the length of the extrudate. That is caused by the fact that the deformation along the diametrical direction was restricted by tooling, while the material along the length direction underwent free deformation. As shown in Figure 7.43b, H_1 represents the length of the inner part of the second extrudate, and the value of H_1 is greater than the designed value due to the effect of additional material flow in the blanking operation. The heights H_2 and H_3 of the extrudates decrease with grain size, which is related with the different material flow patterns under different annealing temperatures. The height H_4 represents the length after extrusion processes, which decreases with the annealing temperature. In addition, the shape stability of the formed part deteriorates with the increase of grain size due to the inhomogeneous deformation with coarse-grained material. Therefore, the length of the extrudate is interactively affected by material properties, die parameters, workpiece thickness, and the compliance of the progressive forming system. In the development of the miniaturized part for a given material, the desired length of the extrudate can be obtained accurately via compensating the punch stroke and applying the fine-grained material.

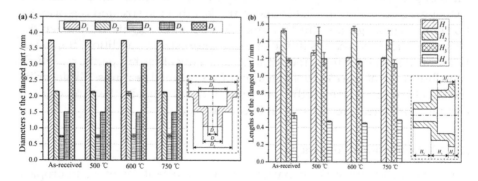

FIGURE 7.43
Dimensional variations of the flanged part: (a) diamaters and (b) lengths [17].

7.5 Summary

In this chapter, the meso- and microscaled bulk forming of sheet metals to produce bulk parts and components via progressive and compound meso- and microforming of sheet metals are extensively presented and articulated. The process is promising and has a good potential for wide applications in making meso- and microscaled bulk metal parts directly using sheet metals. The process characteristics and uniqueness are delineated, and the detailed deformation behaviors and process performance are systematically described. From dimensional accuracy perceptive, the issues and difficulty of the accurate dimension assurance and control induced by the scattering behaviors and performance are analyzed. By using ductile fracture criteria, the ductile fracture behaviors are analyzed. By employing case study parts and its fabrication, the process, tooling, fabricated part quality, and dimensional accuracy involved in this unique and promising process are presented and analyzed.

References

1. M.W. Fu, W.L. Chan, A review on the state-of-the-art microforming technologies, *International Journal of Advanced Manufacturing Technology* 67 (2013) 2411–2437.
2. M.W. Fu, J.L. Wang, A.M. Korsunsky, A review of geometrical and microstructural size effects in micro-scale deformation processing of metallic alloy components, *International Journal of Machine Tools and Manufacture* 109 (2016) 94–125.
3. T. Masuzawa, State of the art of micromachining, *CIRP Annals – Manufacturing Technology* 49(2) (2000) 473–488.
4. J. Giboz, T. Copponnex, P. Mele, Microinjection molding of thermoplastic polymers: Morphological comparison with conventional injection molding, *Journal of Micromechanics and Microengineering* 19(2) (2009) 025023.
5. J. Giboz, T. Copponnex, P. Mele, Microinjection molding of thermoplastic polymers: A review, *Journal of Micromechanics and Microengineering* 17(6) (2007) R96–R109.
6. R. Zauner, Micro powder injection moulding, *Microelectronic Engineering* 83(4–9) (2006) 1442–1444.
7. B. Tay, N.H. Loh, S.B. Tor, F.L. Ng, G. Fu, X.H. Lu, Characterisation of micro gears produced by micro powder injection moulding, *Powder Technology* 188(3) (2009) 179–182.
8. V. Piotter, T. Benzler, T. Gietzelt, R. Ruprecht, J. Hausselt, Micro powder injection molding, *Advanced Engineering Materials* 2(10) (2000) 639–642.
9. F. Vollertsen, H.S. Niehoff, Z. Hu, State of the art in micro forming, *International Journal of Machine Tools & Manufacture* 46(11) (2006) 1172–1179.
10. M. Geiger, M. Kleiner, R. Eckstein, N. Tiesler, U. Engel, Microforming, *CIRP Annals – Manufacturing Technology* 50(2) (2001) 445–462.

11. M.W. Fu, W.L. Chan, *Micro-scaled Products Development via Microforming: Deformation Behaviors, Processes, Tooling and Its Realization*, Springer, London, 2014.

12. A. Diehl, U. Engel, M. Geiger, Influence of microstructure on the mechanical properties and the forming behavior of very thin metal foils, *International Journal of Advanced Manufacturing Technology* 47 (2010) 53–61.

13. A. Diehl, U. Engel, M. Geiger, Mechanical properties and bending behavior of metal foils, *Proceedings of the Institution of Mechanical Engineers Part B Journal of Engineering Manufacture* 222 (2008) 83–89.

14. M. Arentoft, R.S. Eriksen, H.N. Hansen, N.A. Palden, Towards the first generation micro bulk forming system, *CIRP Annals – Manufacturing Technology* 60 (2011) 335–338.

15. M.W. Fu, W.L. Chan, Micro-scaled progressive forming of bulk micropart via directly using sheet metals, *Materials & Design* 49 (2013) 774–783.

16. B. Meng, M.W. Fu, C.M. Fu, J.L. Wang, Multivariable analysis of micro shearing process customized for progressive forming of micro-parts, *Materials & Design* 93 (2015) 191–203.

17. B. Meng, M.W. Fu, C.M. Fu, K.S. Chen, Ductile fracture and deformation behavior in progressive microforming, *Materials & Design* 83 (2015) 14–25.

18. Y.W. Stegeman, A.M. Goijaerts, D. Brokken, W.A.M. Brekelmans, L.E. Govaert, F.P.T. Baaijens, An experimental and numerical study of a planar blanking process, *Journal of Materials Processing Technology* 87 (1999) 266–276.

19. M.D. Gram, R.H. Wagoner, Fineblanking of high strength steels: Control of material properties for tool life, *Journal of Materials Processing Technology* 211 (2011) 717–728.

20. W.L. Chan, M.W. Fu. Meso-scaled progressive forming of bulk cylindrical and flanged parts using sheet metal, *Materials & Design* 43 (2013) 249–257.

21. K. Mori, Y. Abe, Y. Kidoma, P. Kadarno, Slight clearance punching of ultra-high strength steel sheets using punch having small round edge, *International Journal of Machine Tools and Manufacture* 5 (2013) 41–46.

22. E. Ghassemali, A.E.W. Jarfors, M.J. Tan, S.C.V. Lim, Dead-zone formation and micro-pin properties in progressive microforming process, *Steel Research International* (Special Edition) (2011) 1014–1019.

23. M. Zhan, C.G. Gu, Z.Q. Jiang, L.J. Hu, H. Yang, Application of ductile fracture criteria in spin-forming and tube-bending processes, *Computational Materials Science* 47 (2009) 353–365.

24. M.G. Cockroft, D.J. Latham, Ductile and workability of metals, *Journal of the Southern African Institute* 96 (1968) 33–39.

25. S.I. Oh, C.C. Chen, S. Kobayashi, Ductile fracture in axisymmetric extrusion and drawing-part 2: Workability in extrusion and drawing, *Journal of Manufacturing Science and Engineering* 101 (1979) 36–44.

26. P. Brozzo, B. Deluca, R. Rendina, A new method for the prediction of formability limits of metal sheets, in *Proceedings of the 7th Biennial Congress IDDRG (International Deep Drawing Research Group) on Sheet Metal Forming and Formability*, Amsterdam, October 9–13, 1972, pp. 9–13.

27. T. Ayada, T. Higashino, K. Mori, Central bursting in extrusion of inhomogeneous materials, *Advanced Technology of Plasticity* 1 (1987) 553–558.

28. J.R. Rice, D.M. Tracey, On the ductile enlargement of voids in triaxial stress fields, *Journal of the Mechanics and Physics of Solids* 17 (1969) 201–217.

Index

Milton Keynes UK
Ingram Content Group UK Ltd.
UKHW021620071024
449327UK00020BA/1132

9 780367 571276